360 Degree
Spherical Video

360 Degree
Spherical Video

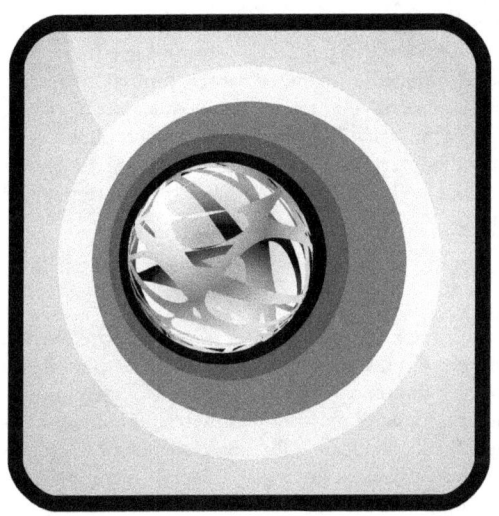

The Complete Guide
to 360-Degree Video

John J. Hussar

Grey Goose Graphics
Endwell, New York

360 Degree Spherical Video Book

John J. Hussar

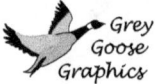 Grey Goose Graphics, LLC
633 Valleyview Drive
Endwell, NY 13760

Find us online: www.sphericalvideobook.com / www.GreyGooseGraphics.com

Printed in the United States of America

Publisher's Cataloging-in-Publication Data

Names:	Hussar, John J., author.
Title:	360 degree spherical video : the complete guide to 360-degree video / John J. Hussar.
Description:	First edition. | Endwell, NY : Grey Goose Graphics, [2016] | Includes index.
Identifiers:	ISBN: 978-0-9983760-0-4 (print) | 978-0-9983760-1-1 (ebook) | LCCN: 2016919294
Subjects:	LCSH: Video recording. | Three-dimensional imaging. | 3-D video (Three-dimensional imaging) | Virtual reality. | Photography, Panoramic. | Photography--Digital techniques. | Image processing--Digital techniques. | BISAC: PHOTOGRAPHY / Techniques / Cinematography & Videography. | PHOTOGRAPHY / Techniques / Equipment.
Classification:	LCC: TR860 .H87 2016 | DDC: 777.65--dc23

First Edition

This book is dedicated to
my wife MaryLou and my daughter Allie,
who have both provided me their support,
guidance, and patience
not only during this project
but also on our life journey.

Contents

Expanded Contents

Expanded Contents

Preface

In the fall of 2016 my team found ourselves in quite a predicament. One of our newer clients contacted us about an upcoming video production project. While it had been scheduled for several months, they asked us to incorporate 360-degree video into the project with less than thirty days of lead time from the scheduled shoot. We had seen 360-degree video technology on social media, and we had seen examples of this type footage as part of a news story on television, so we figured, "How hard could it be?" We immediately acknowledged their request and assured them we would make it happen.

Like many studios in our industry we are always pushing the limits with our production team, our equipment, and our capabilities. We have obtained the real-world experience that has allowed us to become one of the fastest-growing studios in our region. It was (and still is) not uncommon for us to get a special request and accommodate it by adding some gear to one of our production vehicles, learning a new technique, or developing a new skill set. Over the years we had developed an extensive network with other leaders in our industry, key equipment vendors, and similar studios who share the same passion for success. We have several resources to utilize and leverage when we need help, get stuck, or are looking for something beyond our current capabilities.

We recognized immediately that 360-degree video was something we had not yet done. We needed help. We turned to our network and found that none of our partners were using this technology. We turned to our suppliers and found that many of them had planned on distributing this equipment, but it was not yet available. We spoke with several manufacturers, but they only provided us with the anticipated release dates of their gear. We then turned to the Web. We visited several manufacturers' websites, read several prerelease reviews and blog posts, and watched some examples of the technology in use. We played with a few of the established video platforms as well as emerging ones. In just a few hours it became apparent that not only was this technology evolving before our eyes, but many of the numerous questions we had were not yet addressed anywhere. There was not a single source addressing the use and integration of 360-degree video technology. The idea for this book was born.

Over the next several weeks, we read everything we could find about this new technology. We quickly made friends with the vendors selling cameras and the delivery drivers who were making almost daily trips to our studio. We purchased and experimented with several camera platforms. Most importantly, we shot a lot of video. We shot in many settings and under a variety of lighting conditions, and we experimented with different placements of the cameras during single- and multiple-camera shoots. We made lots of notes, took numerous measurements, and reviewed hours of footage. We experimented with post-production workflows, exported to a variety of platforms, and viewed our work on various devices. We then combined all this information and assembled this reference book with the intent to provide a comprehensive guide for the amateur or professional videographer.

It's important to note that this book is not just about the technology, but about how to use the technology, how to create or improve a video production work flow with this technology, and how to create a revenue stream. Readers will also notice that some of the information in this book can be applied not only to 360-degree video but to all video platforms and production enterprises.

This is the first book I've written using our team. We've had the pleasure of working together at Grey Goose Graphics for many years. With an international client base, our real-life experience starting with 16 mm film, Super 8 film, and DVC Pro Tapes and going on to the digital equipment of today has provided us with a solid foundation for this book. As early adopters of technology in this dynamic industry, we work with multimedia using a variety of platforms daily. We have successfully developed numerous marketing campaigns and mastered the use of social media. The fact that our business model is always evolving to keep up with the needs of our clients and the always changing technology in our industry has made us the envy of our competitors. We believe the integration of 360-degree video into our studio offerings may be one of the most significant additions since we began exceeding client expectations when our studio opened, and we are very excited to share our findings with our colleagues.

We successfully completed the shoot our client requested. In fact, some of the early footage we obtained on their shoot became the cornerstone for the research we completed while compiling the data for this book. Additional research was obtained through our relationships with the many organizations and nonprofits within our community. We are proud to be active in our respective communities outside of the studio. Networking is very essential. But we believe that volunteering for worthy causes and supporting nonprofit organizations is not only a way to give back to the community that has supported us on our journey to success but also a way to positively impact many around us. The benefits of our individual and team contributions to those around us has been positive for us both personally and professionally. In this case, we could get access to locations, situations, and events in a very short time interval. We would encourage all our readers to get involved in their communities. This has been an important part of our success.

I would like to formally thank our crew and their families for their support and encouragement during this process. While we may have missed a few engagements and didn't get to spend as much time as normal enjoying the outdoors with them while writing this book, we believe they are happy to be an important part of our success. If it were not for our families, we would have lost our motivation and drive to succeed many years ago.

We hope you find this book helpful and relevant to your endeavors. We enjoy connecting with our readers and invite you to stay in touch.

Visit us on line at: www.SphericalVideoBook.com

Follow us on Facebook at:
www.Facebook.com/SphericalVideoBook.com

Email us at: author@SphericalVideoBook.com

Looking for some help? We are always up for drinking a pint of cold craft beer with you as we contemplate our next project. Contact us today!

Disclaimer

For many years, our studio has strived tirelessly to stay ahead of the pack in our offerings, technology, and work ethic. The intentions of this book are threefold:

1. To share our experiences to allow others to learn this technology quickly. It took us a long time to experiment, develop a working knowledge of the technology, and integrate it into our studio offerings and workflow. We hope this will speed up the process for others.
2. To provide some suggestions and present some of our real-life experience to prevent others from experiencing some of the pitfalls, production delays, and lost revenue we experienced over the last ten years.
3. To offer some suggestions, ideas, or concepts of what can be done when things go wrong with planning, during a shoot, or in a postproduction workflow.

We believe this book will offer both serious hobbyists and professional videographers a comprehensive guide to using 360-degree video. It's important to understand that it's only a guide.

It's meant to offer ideas, suggestions, and advice based on our research and our experience. It's not a guaranteed solution, because what was successful for us may not be right for you.

This guide does not replace common sense, sound judgement, or any rule, regulation, or statute that you may need to conform to.

If you are a professional or even an amateur recording something special for a special someone, it is our hope that you will read this book from cover to cover (or at least all of chapter 9) before you head out to shoot a video. There are lots of examples of things that may help you along the way and provide you with some guidance on both what and what not to do.

While we consider ourselves to be very responsible individuals, there is no way we can be responsible for each of our readers. You are ultimately responsible for your successes as well as things that don't go well.

We welcome you to contact us at any time with any feedback you may have about this book, with any questions you may have, or if we can help you along the way.

Acknowledgments

I would like to acknowledge the following people who have helped make this book possible:

Lori, Warren, and Jorge, for helping me make sense of this all, but more importantly helping me present this information so others could understand it.

Mike, for his continued guidance and technical support along the way.

Chris, For the many hours of discussions, reality checks, and wake-up calls that have kept this project on the road to success.

John, for planting the seed …who would have thought that saying "You should write a book." would have resulted in this?

Jeff, Sam, Zac and the entire Grey Goose Graphics crew, for stepping up in many ways—from the technical review assistance with some of the design elements to covering some of my day-to-day responsibilities while I was working on this project. I couldn't have pulled this off without all of you.

And last but certainly not least, my parents John and Nancy, who have pushed me from an early age and helped shape me into an overachiever, which has led to my success in life.

Chapter 1
Introduction

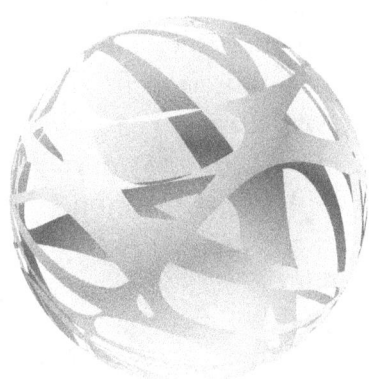

The advance of consumer technology and the availability of affordable devices continue to accelerate year after year. Computing power that once required hundreds of square feet to house has been surpassed many times over by the powerful chips within most handheld devices. Audio signals, once transmitted by towers and processed by transistors in radios powered by alkaline batteries, have been replaced by satellite transmissions or web broadcasting and processed by computer chip technology in a variety of devices. Just about everything we interact with daily has been influenced or changed by technological advances.

360-degree video, a recent advance in video acquisition and playback, is considered by many to be the last step in the evolution of video. This cutting-edge technology is the cross or combination of two technologies many are familiar with—video and panoramic photography.

The History of Video

Nothing makes for a better story than having something visible to further convey the storytellers' point of view, validate their facts, or perhaps illustrate an idea. From early cave drawings to the murals of 20 AD in ancient Rome, the use of such illustrations provided not only something visible but also a way to communicate that could be shared long after the storytelling was over. Throughout time, these storytellers and their audiences provided the inspiration for continued changes and improvements in the way stories are told.

As far back as the mid-1600s, the storytellers were experimenting with the first motion pictures. Images were scored on glass slides, and the slides were moved on top of each other to convey the feeling of movement within the compound image. These devices were known as "magic lanterns."

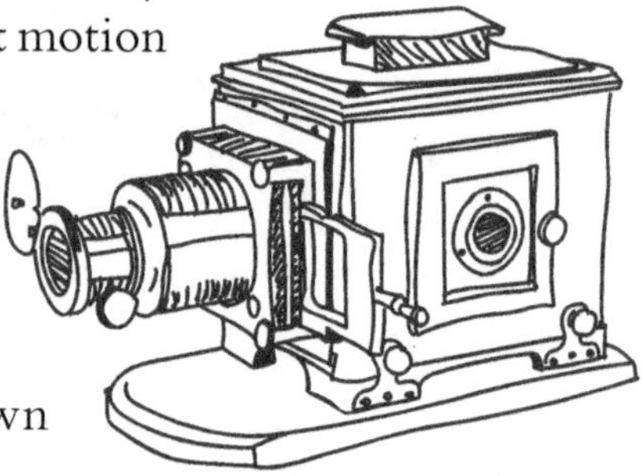

During the latter part of the 1800s, storytellers could take advantage of the many advances offered by film.

Figure 1.1

Magic lanterns provided storytellers of the mid-1600s with the first motion pictures.

In the mid-1800s, illustrators used sequences of several drawings and photos that, when combined and rapidly flipped or fanned, resulted in the appearance of movement within the image. These techniques were the foundation for frame-by-frame video and animation that was eventually expanded on many years later.

Early motion picture cameras capable of recording images to film were soon developed, and in the 1880s the use of film to record these motion pictures began. Images or events were recorded and then projected from the film to audiences. Early on there was no audio with these projections, which explains why these are called "silent films." Instead, a pianist, organist, or even a small orchestra would often play music to accompany the films. In 1927, synchronized audio tracks were introduced with motion pictures.

For over fifty years, film was used as the medium to record and project motion pictures. From the recording of war-time news to the early productions aimed at entertaining audiences around the globe, motion pictures became the new medium for storytellers and audiences of all ages. While motion pictures or films were very popular and highly desirable, they had a few significant limitations. In addition to the expense needed to shoot with film, develop the film, and edit it for the final production, the interval from the time of acquisition until the public could see it would take at least several days, but more often several months. Some of the larger movie productions took years to shoot, edit, and produce.

Many inventors in the late 1800s and early 1900s made significant advances in developing the ability to transmit and receive a signal with data that could be used for communication. Television emerged with the first transmission of a moving image in 1924, the first transmission of a human face in 1925, and the ability to deliver live content in 1926. Many consider this to be the birth of video.

From those days through the present, there have been continual advances in the quality of the images and audio transmitted from storytellers to their audiences.

Panoramic Photography meets Video

In 1843 the first panoramic photos were produced, giving storytellers the ability to capture several images that when aligned provided a greater field of vision for the audience. In 1888, the development of flexible film allowed developers processing photos to combine many images into a single print.

Over the next 150 years, a variety of technological advances improved not only the quality and clarity of panoramic photography but also made acquisition significantly easier for the photographer. Advances in lens technology and the introduction of

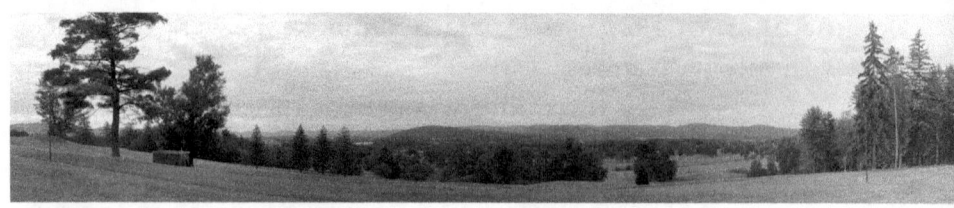

Figure 1.2
Combining several single photos to create a panoramic photo dates back to the mid-1800s.

affordable digital photography in the 1990s, combined with the development of software for personal computers, introduced a new method commonly referred to as digital stitching. The use of stitching software allowed photographers to take several photos of a subject and "stitch" them together to form a larger image, often a panoramic photo. Continued advances in stitching technology and computer processors over the past few years have put stitching software on many cell phones and handheld devices, allowing an amateur user of one of these devices the ability to create stunning results with minimal (if any) photography experience.

The progression from panoramic photography to panoramic video was the next logical step. For many years, military and commercial simulation companies have been emulating a panoramic video experience. Early renditions used a series of synchronized video monitors aligned in rows to give the user the sensation of a virtual reality. Today—because of the continued evolution of wide screen and curved video displays and improvements in the speed and capabilities of computer processing, along with the video stitching of the visual effects—the panoramic video experience is incredible. When this video is combined with advanced sensor technology that allows the user to interact with or control the image being displayed, the resulting virtual reality systems are nothing short of amazing.

360-degree video is a combination of all this technology that allows a videographer to obtain video in a 360-degree sphere around the camera. The resulting video footage may be used in record/playback and/or live streaming platforms. This

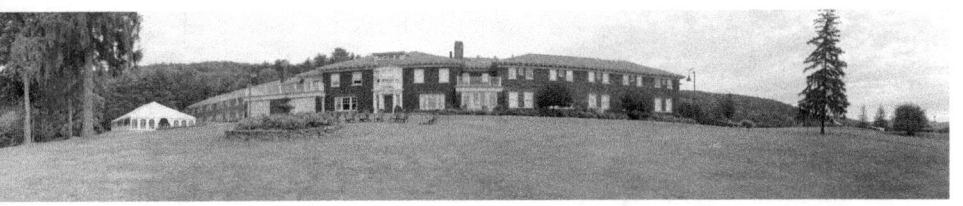

technology is heavily reliant on the advances in cameras, video data management, and distribution platforms.

The Evolution of the Video Camera, File Format, and Storage Media

Video remained a "live" technology until 1937, when it was first recorded to an electronic medium. In 1951 the first video tape was recorded; however, tapes were not available for sale until 1956. From 1956 until 1971 this technology was only available to commercial entities or the very wealthy, as video recorders cost tens of thousands of dollars each. In 1971, SONY released the first videocassette recorder (VCR) for consumers. This recording and playback technology was quickly accepted and made popular by consumers from many walks of life. Over the next several years, VCRs became mainstream in many countries. The technology and the formats of the recording tape evolved. VHS became the standard format and remained popular for many years. During the 1990s, new technology emerged that allowed traditional analog recordings to be stored in a digital format. This new format not only improved the storage capacity of these tapes but significantly improved the quality of the video and audio being recorded to them. The digital format also facilitated the use of a variety of digital devices and computers for the recording, playback, and editing of this medium.

Although digital and analog video tapes remained a common media for many years, video tapes had a few drawbacks. Video tapes

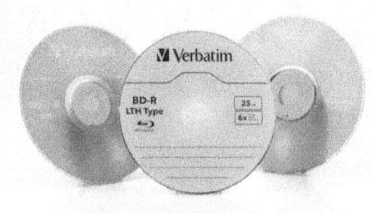

Figure 1.3

The type of disc used in the distribution of a video will determine the resolution of the video presented on the disc.

wore out with use, and they were sensitive to heat, sunlight, and moisture or humidity. The tapes were eventually replaced with the plastic discs. Known as digital video discs or digital versatile discs (DVDs), they are significantly more durable and not affected by light, moisture, or humidity, though they still are negatively affected by heat. Discs also bring a new issue, for they are produced in many formats and various sizes, and some can be only written to in a single session, whereas others may be rewritten many times. Not all DVDs are compatible with all DVD players. DVDs also have a maximum resolution of 720 pixels by 480 pixels.

Further advances in the development of higher-resolution televisions and monitors created an opportunity for higher-resolution playback devices and discs. Blu-ray discs were introduced to the market in the early 2000s and offered consumers a disc with properties and a physical size like a DVD; however, they store a much larger volume of data and offer a maximum resolution of 1920 pixels by 1080 pixels.

The newest format, which is just emerging as of the writing of this book, is Ultra HD Blu-ray discs, which provide a larger volume of data over a standard Blu-ray and yet a higher resolution of 3840 pixels by 2160 pixels.

This resolution is commonly referred to as 4K. Many experts believe that this resolution is the highest that can be perceived by the human eye, so further advances may be considered pointless by many.

As many advances were made in the playback and playback storage technology, cameras, recording technology, and storage devices have also significantly evolved. Early video cameras could weigh several hundred pounds and thus were often stationary on top of a tripod within a studio. Over time, because of technological advances, the physical size and weight of the cameras continued to become smaller and lighter, leading to the portable camcorders of the 1980s and 1990s. These camcorders usually incorporated a shoulder rig, allowing users to carry or rest the camera on their shoulder when it was not being used on a tripod.

With the integration of digital technology and (eventually) the elimination of tape media, the motorized mechanisms and recording heads were replaced with digital readers. Replacing the moving parts with computer chips and digital media also made the cameras more efficient, which reduced the need for larger, heavier batteries. This allowed for significant reduction in the weight and size of the camera, which led to the release of several handheld camera options.

One of the more popular handheld devices used by amateurs and professionals alike is the DSLR camera. DSLR is an abbreviation for digital single-lens reflex. A DSLR is a digital camera that combines the lenses, optics, and the mechanisms of a single-lens reflex camera with a digital imaging sensor. While original models were only capable of still photography, most current DSLR cameras can obtain high-quality video footage. In fact, DSLR cameras are our camera of choice, as they combine the ability to incorporate a variety of lenses with a camera body that gives us a very flexible acquisition platform without an extremely significant capital investment.

With the continual advances made in smartphones and tablets, cameras capable of acquiring both still images and video have become a standard feature of these devices, and this feature sometimes may be the deciding factor when users are selecting one of these devices for purchase. The quality and capabilities of these devices have advanced to the point that many people who once carried still cameras, video cameras, and cell phones to special events or on vacation are now simply taking photos and videos with just their smartphones.

Despite the advances in the handheld and amateur equipment areas, serious photographers and/or videographers continue to purchase high-end equipment with additional functionality or other specialty cameras designed to acquire images and video in a specific situation or environment. One clear example of this is recent product launches for action or extreme sports situations and those integrated with drone technology.

One of the most recently released categories of cameras are those capable of obtaining 360-degree video. It's important to understand that these cameras are made possible by advances in four specific areas.

Lenses. Significant improvements have been made in lens technology. Lenses today are smaller, have better quality, and (because of new materials) are less expensive than they were just a few years ago. These advances can be combined with the development of the digital zoom, which uses sensor technology and processors to magnify images without the need for the weight and expense of traditional glass zoom lenses.

Sensors and Processors. The sensors within digital devices have also significantly improved. The resolution of the sensors continues to improve, allowing the subject matter to be replicated more effectively. With continued improvements in microchip technology, the data from these sensors is processed more quickly and efficiently.

Data Storage. Storage device technology continues to improve, and we continue to place more and more data on smaller and smaller cards. Released about 1984, the original data storage cards were known as CompactFlash cards. These cards measure 43 mm × 36 mm × 3.3 mm and could store approximately 512 GB of data. Throughout the 1990s, additional formats were released, including multimedia cards and early versions of the secured digital cards (SD). These new cards offered a smaller form factor, measuring 24 mm × 32 mm × 2.1mm, and provided storage capacity ranging from 4 to 16 GB. About 2007, the microSD card was introduced, which had a much smaller footprint measuring only

11 mm × 15 mm × 1 mm, but a greater capacity and transfer speed. Over time, the footprints of these cards have remained constant; however, we have seen significant increases in not only the amount of data capacities of these cards but also the transfer rates. Cards today have a transfer rate of up to 100 times faster than when they were originally released and have a total storage capacity of up to 2 TB.

Power. As sensors, processors, and storage technology improved, there has been a significant improvement in the overall power consumption of cameras. This, in combination with improvements in battery storage, charging, and storage technology, led to a noticeable decrease in the size and weight of the batteries, which continues to help reduce the overall weight and size of cameras.

Continual advances in these areas will not only lead to additional capabilities of 360-degree digital video but also bring down the cost of this technology, making it more affordable and accessible to many.

Video on the Web

There has been no greater impetus for increases in video on the web than the purchase of YouTube by Google. Many thought leaders in the industry credit this platform for the expansion of the video industry, for it provides a mainstream ability for people from all over the world to communicate through this medium. YouTube has evolved since the acquisition and remains the leader in streaming capabilities. Other companies such as Vimeo and Facebook continue to chase Google. There are also several other proprietary platforms, streaming services, and sponsored distribution channels available.

Google and Facebook are currently capable of streaming 360-degree video. Several camera manufacturers have also developed their own apps or platforms to stream video from their specific devices. There is no doubt that the numbers and

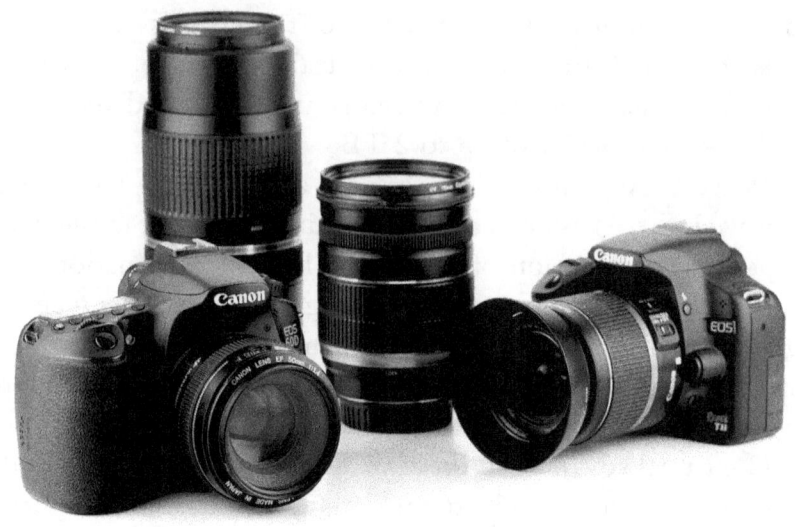

Figure 1.4
DSLR cameras offer videographers a flexible platform for acquiring traditional video without requiring a significant capital investment.

capabilities of these distribution channels will continue to evolve and grow.

360-Degree Consumer and Professional Video

Affordable or accessible 360-degree video has been made possible by the evolution of the video camera, stitching technology, storage solutions, and delivery platforms. For the remainder of this book, we will be focusing on 360-degree video technology. We believe the integration of this new technology into your offerings and workflow has almost endless possibilities. With the advance of technology and the resulting affordability of this platform, we think that by the end of this book you will agree that it's something you cannot simply overlook or ignore.

Chapter 2

360-Degree Video Overview and Integration

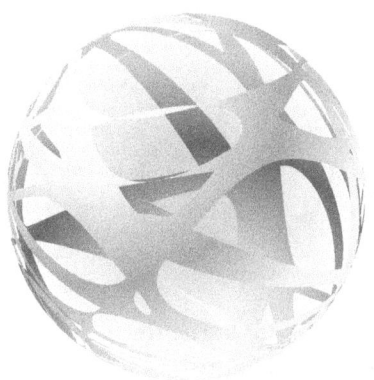

3 60-degree video is a technology that combines mediums and is shot using cameras made of several advanced components that have evolved over many years.

What Is 360-Degree Video?

In the simplest definition, 360-degree video is a compilation of video shot by multiple lenses and stitched together to provide the user a 360-degree perspective from the point of view of the camera.

Currently this evolving technology is referred to by many other names, including spherical video, 3D video, and virtual reality.

Depending on the actual footage taken, the acquisition devices, the playback platform, and the playback devices, the viewer may also interact with the creation of the video by changing their perspective, field of view, or position within the video.

11

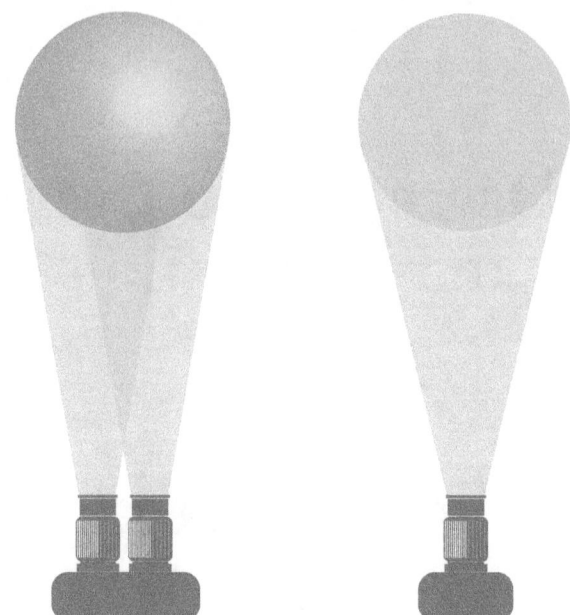

Figure 2.1
Steroscopic video is required for a true three-dimensional video experience. The problem is that distribution options remain very limited.

The video may be monoscopic or stereoscopic. Monoscopic video is flat, and the same image is displayed to viewers for each eye. Stereoscopic video uses at least two lenses to capture unique footage for the left and right eye, which provides depth to the video but also requires a stereoscopic device and/or platform for viewing. For the remainder of this book, we will focus on monoscopic 360-degree video.

How Does 360-Degree Video Work?

360-degree video is captured like normal video; however, the images are stitched to create a spherical field of view. Early versions of this technology used several cameras that were specifically aligned and synchronized to capture several different footage segments. These segments were then manipulated and stitched together using a variety of different software packages. As technology evolved, these individual segments began to be processed within the camera.

Both models are still available and are being used today. It's essential for both the amateur and professional to consider these differences as well as understand how the individual camera platform can greatly affect your desired outcome. Please see chapter 3 for more detailed information.

What Can 360-Degree Video Do Well?

360-degree video can tell a story like no other platform. The viewer is instantly placed in the center of the action. With the movement of the mouse, flick of a finger, or on some devices just tilting the device, users can interact with and watch the footage from their own perspectives.

Imagine capturing an outdoor performance by your favorite musical group. On playback, you are right back at the performance, for you see not only the performing group, but also the audience and their interaction, the open sky, and all other activity that was visible at the time of the recording.

Imagine snorkeling on a reef with a 360-degree video camera. As you swim about with the turtles and fish, you are capturing the first-person perspective as you would with any other traditional camera device. But ... have you ever looked behind you? Did you see the movement of the fish after you passed? Did you see the other sea creatures reemerge from their homes within the reef? What about the near miss of the stingray at your feet? 360-degree video provides not only the opportunity to replay the moment in time at a later date, but it may also let you view other things unnoticed at the time of the recording.

Later in this chapter, we will introduce many different ideas for the use of this technology.

What Are the Limitations of 360-Degree Video?

The use and integration of 360-degree video is truly in its infancy. Until recently, this technology was cost prohibitive to most production companies and to all but a very few individuals. It is currently very difficult to cover all the limitations or concerns related to this new technology, as even the very early adopters have very little experience with this platform.

Before we share the limitations discovered during our experimentation and early experience with 360-degree video, it is important to point out that our findings may be different from those of some other early adopters. We also realize that these limitations may vary from platform to platform or from camera to camera. And lastly, as the cameras in this market sector continue to evolve, we expect the manufacturers to address these limitations with continued advances and the release of new cameras.

During our initial experiments and implementation of 360-degree video in our studio offerings, we identified the following issues:

Positioning. Positioning of the camera is significantly different than it is when using any other camera platform. Depending on the different lens configurations and platforms, we found that blind spots may occur within two to three feet of the camera bodies.

These cameras see everything in the 360° sphere. Think about your typical production and everything that currently goes on behind your lens. What if you are using this as part of a multi-camera array, or perhaps you are using two 360-degree video cameras? What do you want in your shot? What do you need to change, consider, or reposition?

Lens and Sensor Capabilities. It's important to note that not all 360-degree video cameras shoot a full 360-degree sphere. Some camera configurations are limited to producing 240-degree or 270-degree spheres.

Not all camera rigs have the same lens or sensor configurations, and not all sensors are created equal. This is most easily explained by the following example.

Consider placing your 360-degree video camera in a classroom in which a presentation is occurring and where an LCD projector for slides is also in use. If you are using a camera configuration in which the images are obtained by multiple

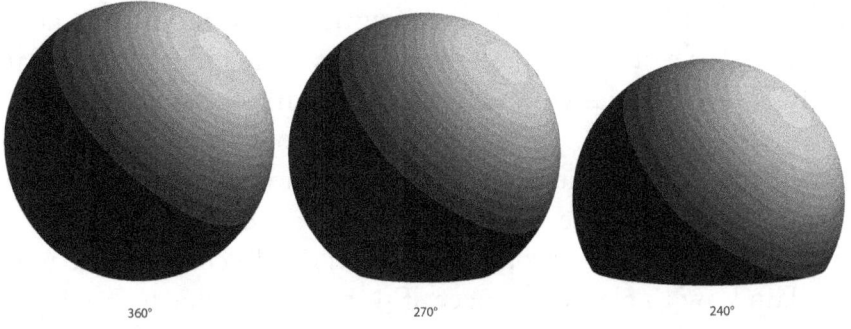

Figure 2.2
Not all cameras shoot a full 360-degree sphere. Consider the camera's capabilities when using a camera on a shoot.

cameras with a smaller field of view instead of fewer cameras with lenses that have a larger field of view, your video will most likely be considerably improved. Each lens having its own sensor and focusing on a smaller area allows the system to adjust for more drastic lighting or exposure changes within the sphere. Using a rig with fewer lenses or sensors forces the smaller number of sensors to detect and adjust to a larger portion of the sphere, which may produce significantly different results.

Audio. In general, we found the onboard audio of most the 360-degree camera platforms to be lacking. Like other cameras with onboard audio, the quality of the audio produced may be sufficient for quick streaming or possibly personal use, but we did not find it worthy of inclusion in professional productions. Quality audio acquisition requires additional planning and is discussed further in chapter 6.

Cost. During our initial research and ultimately our purchase of camera platforms for our productions team, we found cost was a significant barrier early on. Initially, the professional-grade cameras of choice cost more than $60,000. We also found several consumer-grade camera solutions with mediocre quality for just a few hundred dollars. Finding the balance between quality and affordability was a little challenging. While we have worked hard to not endorse or recommend any specific camera or platform within this book, it is now possible for a studio to acquire cameras for under $1000 that will have the quality and features required for professional production.

Despite any limitations of this technology, the expanded experience of the viewer is well worth the required consider-ations or changes to a production workflow. In chapter 5 we will explain how to plan for a successful shoot with this technology.

Using 360-Degree Video in the Real World

The use of 360-degree video in the real world can be looked at in many ways. For the hobbyist, basic consumer video may simply be a way of telling a story or documenting an event. For professionals, their footage may be used for marketing, promo-tional, or other business ventures. 360-degree video offers so much more than the traditional mainstream video that is being used today.

Any video produced must be interesting, engaging, and rel-evant to the viewer to get a favorable response. If you do not satisfy any one of these components, the chances of your video successfully creating the desired outcome quickly diminish. Creating an effective video requires the videographer to have a solid understanding and working knowledge of the basics, such as production planning, appropriate focus on a subject, and lighting and framing, as well as knowing how to record quality audio in a variety of environments. All these basic subjects need to be included in your planning for each shoot, and we will spend more time on planning in chapter 5.

Once you have mastered the basics, it is time to get out and shoot. Practice makes perfect, so shoot as much as you can, as often as you can. You may already have several ideas of how to leverage the advantages of 360-degree video. If you are looking for inspiration or ideas about how to incorporate this technol-ogy into your studio's offerings, below is a short list of examples of projects or ways which 360-degree video maybe used or inte-grated. It is by no means a comprehensive list.

Areas for Using 360-Degree Video

Personal

- Document a family event (such as weddings, birthday parties, or funerals)
- Record an adventure or activity (such as hiking, kayaking, or swimming)
- Capture a sporting event or participation in extreme sports (such as ziplining, skydiving, or skiing)

Marketing and Promotional

- Feature a product (for example, display the interior of a vehicle or a product showroom)
- Promote a company (production lines, trade shows, or events)
- Provide a virtual tour (show companies, venues, or real estate)

Documentary

- Document an event (such as a sporting event, dance recital, meeting, or public hearing)
- Create a demonstration (a "how-to" video)
- Physical site surveys (such as property inspections or crime scene documentation)

Education

- Conferences
- eLearning
- Courses, labs, and demonstrations

Special Considerations

While producing a quality 360-degree video requires the same planning and basic skill sets as the production of a traditional video, there are two additional items that a videographer should consider when integrating this technology with the real world.

Length of video. The timing of traditional video is relatively simple; present the subject long enough for the viewer to have a desired response. For some videos, a short, rapid sequence may solicit the desired response, while others may require a subject or text to be presented for a longer interval to communicate the message. With 360-degree video, it's significantly more challenging. You want the viewer to have time to interact, "move around," and take it all in. This is great for the viewer who understands the technology, but what about less technologically savvy viewers who do not realize they can interact with the video? It's a balancing act and requires some additional consideration of your intended audience.

User engagement. For the last several years, the attention spans of technology users have continued to decrease. Websites are designed differently, with fewer clicks required to check out of ecommerce stores. Marketing campaigns have done away with complex messaging and supporting text documents. Videos in general have decreased in length. In general, viewers are looking for more content in a shorter interval. At least that was the case until 360-degree video started to emerge. Early trends are showing that viewers are willing to spend additional time with a 360-degree video, but the video must be engaging or interactive. Let's go back to our previous example of the classroom presentation. If the camera is positioned to focus on the presenter and the slides and viewers start to look around the sphere and see nothing happening, the chance that they will remain engaged is relatively minimal. Now take the same presentation with an interactive speaker. Say he or she is tossing a football around the room and someone catches it in the face, someone falls out of a chair, or perhaps something gets tipped over ... the video just got more interesting. While we certainly do not endorse throwing footballs at people's faces, knocking them out of their chairs, or dumping something on a table, it's essential to show that something is going on in all areas. If not, your production will be no different than a traditional video, except you complicated your production by introducing a more complex platform into the planning and execution.

Chapter 3
Selecting a Camera and Platform

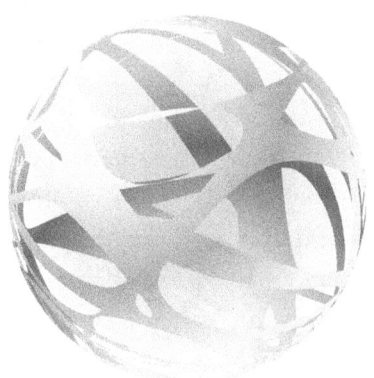

Technology is evolving every day. It's become increasingly harder for the consumer at any level to select cameras. There are many to choose from—not just the extensive product lines and offerings from the long-standing manufacturers, but also the very impressive offerings from new technology companies that are just coming to market. Videographers and production companies should consider not only the intended purpose of the device and studio budget but also the capabilities of the camera and how it is to be integrated into your workflow and ultimately your audience and the requirements of the delivery platforms.

There are several things that affect how consumers purchase products. Consider your most recent trip to the grocery store. Did you have a list? Did you stick to the list? Did you research product nutritional information in one of the aisles? Did you

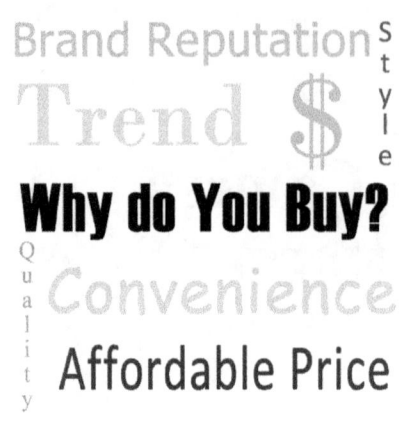

Why do You Buy?

Figure 3.1
Why do you buy?

change your mind based on a display you saw? Did you recall a recent TV commercial? Were you offered a sample of something new to try? Did you get sucked in to an impulse buy while waiting in line at the register? Did you have coupons? Did you purchase a family pack because it was a good deal? Did you buy a name brand or store brand? People think a trip to the grocery store is easy … yet there is a lot to consider!

Most of us watched our parents shop for groceries from a very early age and picked up a few habits. Once on our own, we modified those habits, and after a period of time, we simply headed to the store and got groceries.

With technology purchases, you need to consider everything you do when shopping for groceries, and many more things. There is no right answer for any of these questions. Like buying groceries, many of the answers will be based on what you like, what you've had before, and perhaps some external influence like the product manufacturer, marketing activities, or a friend's recommendation. With a 360–degree camera selection, here are the additional things you need to consider:

Purpose

While there are many things to think about when defining the purpose of this camera purchase, it comes down to this: why are you buying a camera?

- What is the purpose of this camera? (See the discussion of uses in the previous chapter.)
- What is your desired outcome?
- Who is your audience?

- Are you adding a new offering?
- Are you replacing an outdated device?
- Are you looking for a backup device?
- Are you looking to add additional cameras to your shoot or workflow?
- Will this camera be used to train a new member of your team?

Budget

Establishing your budget is the next consideration. Today cameras can be found at all price points, from ones for the basic consumer costing a few hundred dollars to those for the cinematography production house, which may spend tens or even hundreds of thousands of dollars for a camera to meet a specific need, desire, or want. However, that level of investment is not required. It is now possible for a studio to acquire cameras for under $1000 that will have the quality and features required for professional production.

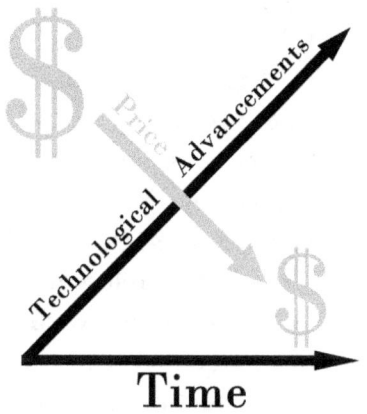

Figure 3.2
Technological advances have made this platform affordable for many.

Capabilities

Here is where the selection of a camera gets harder. There are so many different things to think about regarding the specific features of a camera. Some owners will suggest getting a system as complex as you can afford, while others subscribe to the philosophy of getting just what you need and keeping it simple. You should evaluate as many of your options as possible and make a written list of what you need, what would be nice to have, and what you do not want. Then start your comparison. Here are the things that we considered when purchasing our cameras.

- What are the specifications of the lens?
- Can the camera capture 360 degrees, or is this a rig with multiple cameras that need to have their output manually processed?
- Is the stitching done within the camera?
- What are the frame rates the camera can capture?
- What file formats does it support?
- Can it acquire still photos as well as video?
- What is the compression rate?
- Does it have internal memory?
- Does it support external memory? If so, How much? What kind?
- Does it use an app? What platforms are supported?
- Is it compatible with editing software?
- Does it do well with low light?
- Is it waterproof?
- Is it shockproof/drop proof?
- Is it capable of working in required temperature range? (Does it handle extreme temperatures?)
- What does it use for power?
- Can it be powered via USB?
- How long do the batteries last?
- Can batteries be changed by the user?
- How does it mount?
- What specialty mounts are available?
- Can the image be inverted?
- What is its physical size?
- How much does it weigh?
- Is it drone compatible?
- Does it have a remote?
- What's included in the box?
- What other accessories are available?

Integration Plan

Once the purpose has been clarified, the budget defined, and a comparison of the capabilities completed, the next consideration is to verify either that this new camera will fit your workflow or that your workflow can be reasonably modified to accommodate this camera.

- Does the camera fit into your production vehicle(s)?
- Can the camera be easily shipped to a location?
- Can the camera be checked or carried on a commercial flight?
- Does the camera come with a solid case for travel? If not, are good aftermarket cases available?
- Is it easily packed, set up, and repacked?
- Do you have the required staff with the required skill set to operate the camera?
- Does the camera require a specific technology to operate? (Wi-Fi, Bluetooth, Internet, etc.)
- Will the operation of this camera interfere with other equipment that is already part of your workflow?
- What is required for data transfer from the camera?
- What are the data transfer rates? How long will getting data off the camera take?
- What is required to edit the footage? (Both hardware and software)

Distribution Requirements

Lastly, consider your target audience, perhaps your typical clients, and the distribution platforms you use when buying a camera. Currently several of the leading video distribution platforms either support 360-degree video with specialized controls or have plans to support this technology within the next few months. If you work with specific requirements or must follow the client's directions on distribution, additional research into a distribution plan prior to selecting a camera is advised.

We mentioned earlier the difference between monoscopic and stereoscopic video. While we have focused on monoscopic 360-degree video throughout this book, we would be remiss in our responsibilities if we did not point out that this should be a consideration when selecting a camera.

As this technology continues to evolve, we anticipate that the streaming options for this type of video will also evolve. There are several current options for this streaming technology, and additional releases are anticipated in the coming months. If streaming is an essential requirement, videographers may find themselves having to purchase specialized platforms, as currently there are significant differences in the specifications of available streaming and non-streaming platforms.

Helpful Resources

If you are still stuck after considering your purpose, your budget, the capabilities of the camera, your distribution requirements, and the many resources available, here is our philosophy.

Technology is introduced at a dizzying pace. Updated or improved lenses, optics, and processors are being introduced every day. It is our recommendation that as videographers, you should purchase the most recently released, up-to-date, advanced camera that you can afford. When you consider how fast technology is changing, there is a much greater likelihood of older technology being phased out or surpassed. Buying the latest and greatest will provide the best chance of increasing the longevity of your purchase and maximizing the return on your investment.

Chapter 4
Getting to Know Your Camera

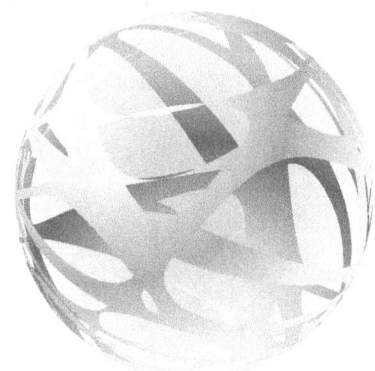

Exploring Your Camera's Capabilities

As with any new device or piece of equipment integrated into your workflow, it is essential that you clearly understand the specifications and working capabilities of your individual camera. While we touched on these briefly in the previous chapter, every videographer should be very familiar with the following six specific camera features before using a camera.

Environmental Tolerances

It is important to understand how environmental conditions and temperature ranges can affect your camera. We have been on shoots where a camera worked very well indoors or in an air-conditioned environment, but when it was moved poolside

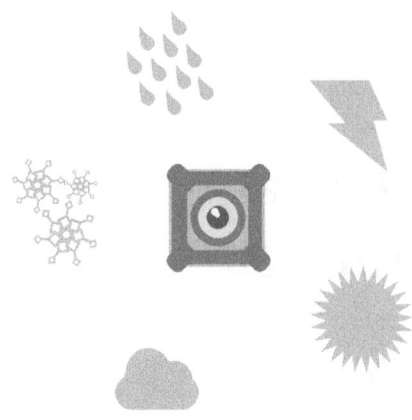

Figure 4.1

While most cameras have high environmental tolerances, it's important that you understand the type of environment in which your camera will function.

and exposed to direct sunlight, the camera temperature quickly exceeded the operating limits, and it shut down to protect the processor and other heat-sensitive components. We have also used cameras when we were very close to the lower temperature limits, and they responded slower and had a noticeable reduction in the clarity of the video. The time to learn this is during tests in the studio, not when setting up for a shoot.

Rapid changes in environmental temperature often result in fogging on the lens and possibly the internal components. If the camera is being used in a waterproof or similar enclosure, the use of anti-fog packs may help reduce lens or enclosure fogging.

It's also important to consider the change in the environmental conditions during a shoot. For example, going from one extreme to the other may create issues not only with the operation of the camera but also with the battery life, which we'll talk about next.

Battery Life and Power Consumption

The battery life of a camera is an important characteristic to understand and identify before any scheduled shoot. Obviously if the battery life of the camera is 45 minutes and you have a 60-minute shoot scheduled, you are going to run out of power. To finish this shoot, you will need a second battery.

Battery life becomes even more crucial if you're running a single camera for a lengthy shoot. Longer shoots require time for the replacement of the initial battery with a secondary battery without missing a key component or specific element of the shoot. For smaller productions where there is a lot of start-and-stop camera action, this is insignificant. When recording or documenting life events or lengthy shoots, this becomes more of a concern and requires either a planned and well-defined stop point at a predesignated interval or the use of multiple cameras and the staggering of power replacements between the multiple cameras at defined intervals.

If you have just purchased a new camera and have not yet purchased a spare battery, this should become a priority before you begin using your camera for anything beyond personal or recreational purposes. We recommend that you always have a spare battery readily available for each camera you are using. Even the best batteries eventually wear out or develop issues. Over the last several years, we have had batteries fail on shoots. From experience, we can tell you that a battery never fails at a "good time" on a shoot. Always be prepared for the eventual failure of a battery.

File Compression and Storage

You also need to understand file compression and storage capacity to be an effective videographer. This may sound obvious, but it's essential that you have enough storage capacity, either on the device or on compatible removable media cards, to cover the entire shoot.

We use removable cards whenever possible. Not only do they offer an almost unlimited shooting time, but they also provide the ability to split the data from the shoot onto multiple storage devices. We'll talk more about why this important in chapter 9.

While many of the newer cameras support very large card storage devices, it is essential to confirm that the card is compatible before inserting it. In addition to verifying that the storage capacity is compatible with the camera, you should also confirm that the card has the proper read and write speed to keep up with the camera. It's important to note that newer cameras are often not compatible with older card media, because many of the older cards do not have fast enough write speeds to keep up with cameras that can quickly process large amounts of data, such as those capable of 360-degree video.

Depending on the capabilities of your camera, you may be able to adjust the storage capacity by adjusting the frame rate, resolution, or compression settings. It's important to note that adjusting any of these settings may result in a change in the overall quality of the footage obtained. There are many tools and resources to compress files, reduce file size, etc. during post-production if editing or delivery size is an issue. Video footage should always be shot using the largest possible format that will still fit on the available storage media.

Lens and Sensor Capabilities

Once you understand what environment the camera will work in and have confirmed that you will have enough battery life and file storage space for your project, it is important to understand the lens and sensory capabilities of the camera.

With traditional video cameras, distance is measured between the subject and the lens. With 360-degree video, it's a bit more complicated. Obviously, we still need to consider the distance between the lens and subject, but now the subject appears in a sphere around the camera. There is no longer a traditional foreground and background. Because everything in the sphere is visible, distance should be thought of as a series of considerations and calculations from the individual lenses. We will go into more detail about this in the exercises later in this chapter.

Different lenses have different resolution capabilities. In the simplest terms, resolution is defined as the pixel dimensions of the video captured. The higher the pixel count, the better the resolution and the acquisition quality of the video. The resolution will not only affect the quality of the video being captured but may also dictate the editing options available in postproduction as well as the quality of the final product. Entry-level products will generally have a lower resolution. Products that are mid-range or developed for the professional almost always have the ability to change the resolution. Unless you have a specific project requirement, as we mentioned previously, it is always recommended to capture video in the format offering the highest resolution that the camera's available memory capacity will allow.

Another variable associated with a camera's sensor is the frame rate. Like resolution, the less expensive cameras will have fewer options in this area in comparison to mid-range or professional cameras. Frame rate is defined by the number of frames per second (fps). In the United States, broadcast television uses 29.95 frames per second. This standard is what most videographers align their camera settings with. Shooting video with a setting that differs from this standard frame rate can produce some great effects, but it requires some advanced editing experience in postproduction. Video acquired at a slower rate, usually 24 fps, will have a "softer," more cinematographical look or feel. Capturing video at a higher frame rate allows the editor to create a slow-motion effect without any loss in resolution. These faster rates may be 60 fps, 120 fps, or greater. The faster the frame rate, the more data is saved. If memory capacity is a concern and you have no intention of using a slow-motion effect, there is no added benefit to shooting a fast frame rate.

One of the most important camera capabilities to understand is the digital sensor's management of light. We all know that too much light will "white out" or diminish the quality of the video. Too little light will darken the video and impact our ability to distinguish specific features or colors.

Most traditional cameras on the market today can make changes to the sensor's interpretation of light by changing white balance, exposure, and ISO settings within the camera. The available settings and features vary from camera to camera, and we recommend that videographers become familiar with the settings available on their specific camera.

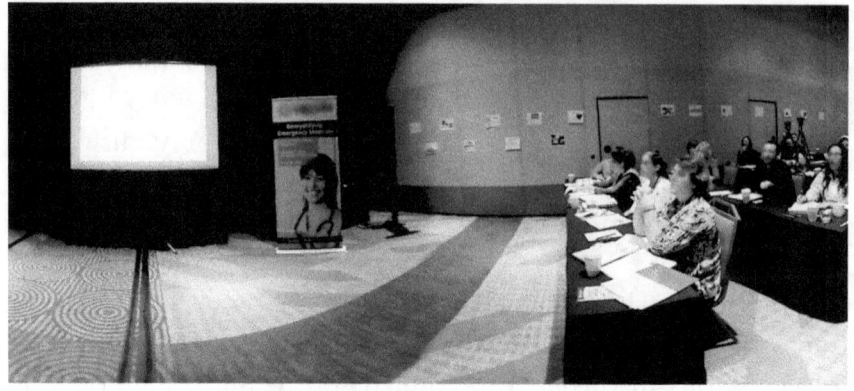

Figure 4.2

Note that the projection screen generates inconsistent lighting in this scene. Unfortunately there are no camera settings that will allow you to alleviate this issue.

Once the settings are defined and recording begins, the camera begins to process the digital data through the sensor. The camera will then sample specific regions or points within the field of view, process all the data by completing a comparison of the sample points to the desired setting, and adjust not only the data that is sampled but all data being recorded and stored.

Let's now consider the 360-degree video platform. While the camera functions slightly differently, the concepts are the same. Video is obtained from multiple lenses and stitched to create a sphere. For most of the cameras on the market that can capture 360-degree video, changing the settings on the camera changes the settings for all lenses or sensors within the camera.

Remember, with 360-degree video we are not only working with multiple lenses, we are also working with light from 360 degrees around the camera. This becomes an issue when lighting is inconsistent within the sphere. Making an adjustment to the camera settings for any area will change the settings for all sensors being used to capture the entire sphere.

At time of the writing of this book, lighting is one of the challenges still being addressed by camera manufacturers. Although many cameras on the market have basic adjustments, we have yet to find a single-camera platform that offers extensive control and management of lighting adjustments for individual lenses at a price point that is affordable for most videographers.

Do not let this discourage you. We'll share some tips and tricks about lighting in the exercises at the end of this chapter.

Mounts

The proper mounting of a camera will not only ensure your success with the shoot but also protect your investment and reduce your exposure for any liability that can result from an accident caused by a camera breaking loose.

A variety of mounts for cameras are shipping with the new platforms. Examples include mounts for tripods, adhesive tape mounts, and wearable mounts (wrist, chest, headband, helmet). Some cameras have standard ¼"-20 tripod mounts. Some require the use of proprietary brackets. It's important that you select the right mount for the project.

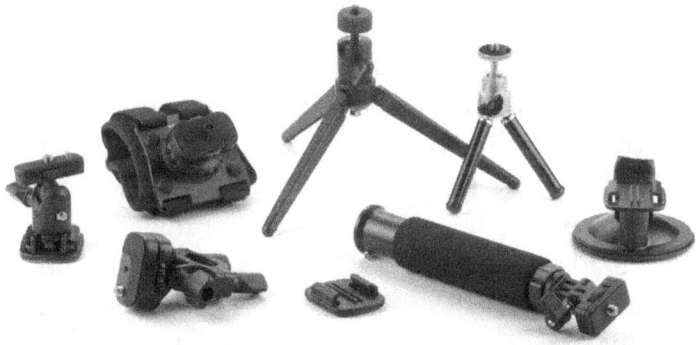

Figure 4.3
There are many mounting options for traditional and 360-degree video cameras.

It's essential that you take the time to learn how to attach your camera to the mount(s) you will use as well has how to affix your mount(s) to the desired object(s). Mounts are discussed more in chapter 5.

Live Streaming

Along the way we have touched on the importance of under-standing your desired distribution platform(s), a topic we will address further in chapters 6, 7, and 8. Live video streaming adds another layer of complexity to any production.

It's important to note that not every camera is capable of streaming video. The time to determine if you have this func-tionality is long before a scheduled shoot that needs it. If your camera has this capability or you are preparing to purchase a camera for this purpose, it is very important that you under-stand streaming-related options or settings such as resolution and frame rate. These will often dictate your streaming platform.

You must also assess the bandwidth or Internet speed that is required for reliable streaming or connecting your camera to the platform. Just a few years ago, when the virtual meeting platforms emerged, we found ourselves assisting clients with video webcast-ing from tradeshows, conferences, and special events. Early on, we were crashing networks in some of the largest conference centers up and down the East Coast by simply setting up a few devices and connecting multiple cameras from a single access point. With faster access points, proprietary compression algorithms, and newer equip-ment, it has become much easier to conduct these webcasts without draining all of a venue's Internet resources, but it still happens. Many rural communities still do not have high-speed Internet access, and many businesses and facilities opt to not pay for this access.

If you are asked to commit to a streaming video job, a site visit and bandwidth evaluation are essential before you make any agreement. If your evaluation is not favorable to the project, you should verify that all identified issues are corrected before making a commitment to provide video streaming services. We strongly advise any videographer not to amend their offerings to provide, subcontract, or coordinate the Internet connection as part of a streaming project. New or temporary Internet access points have a higher rate of failure than established access points. Don't get stuck with a bad situation that you cannot control.

At the time this book was written, identifying compatible streaming platforms was a challenge. Only a couple of mainstream free video platforms were capable of displaying 360-degree video with a fully functional interface. Neither provided an option for live streaming of 360-degree video. There were several proprietary platforms that are either affiliated with specific camera manufacturers or paid services that are testing platforms. We are confident that this will improve with time.

Camera Exercises

Knowing the capabilities and settings of your camera is only part of getting to know your camera. We have developed the following four sequential exercises that we use as part of our internal workflow whenever we add new equipment or members to our video crew. These exercises will force both new and experienced operators to quickly develop a working knowledge of their camera. While some videographers may be tempted to jump to a specific exercise that may be especially relevant to a current project or upcoming shoot, we recommend that all the exercises be at least reviewed if not fully completed in the order they are presented.

Before you get started, you should make sure you have a working 360-degree camera. It needs to have new batteries and be fully charged or powered by an external supply. You also should understand how to turn it on, record video, and turn it off.

Depending on your individual camera platform, you also need to understand how to use the live preview function of your camera or be familiar with how to record and play back your video. Most preview modes shut down as soon as you press record. Should this happen to you, this is normal. To complete these exercises, do not press record.

As we move forward with the exercises, we will focus on the live preview function. If your camera does not have this functionality, you can accomplish the same results by simply recording the exercise, playing back your footage, and making any required adjustments. Please note that you may need to repeat this sequence multiple times before you achieve the desired results.

All exercises are designed for a single videographer; however, working with a team will certainly speed up the process. Although some of the discussion is specific to learning your camera and the 360-degree video platform, much of the information can be used for other platforms and may be incorporated into an everyday work flow.

Exercise 1: Indoor Shoot

At the end of this exercise, you should know:

- How to determine optimal distances for subject matter
- How to determine stitch points and blind spots
- How to check lighting settings
- How to evaluate the camera's microphone

This exercise works well in a large room that has a variety of large stationary objects. We have completed this exercise in many places, including a dining room with a table, the front of a meeting room at a conference center, and the middle of an auto dealership.

In addition to your camera and the ability to preview it, you will need an adjustable mount that will allow you to change the vertical height of the camera. Although every specific location where you shoot will have unique features, the optimal distance or "sweet spot" will remain relatively consistent. For this reason, we also suggest you use a tape measure and record your findings for future use. In fact, we recommend you record your findings for each of these exercises. While the settings may change a little on each subsequent shoot, they will provide a solid starting point to help you plan your shoot. Lastly, we recommend creating an atmosphere with a variety of lighting conditions or equipment to change the lighting. We use standard light boxes with dimmers, an LCD projector with a laptop, an Internet browser, and a few reflectors to evaluate our camera's ability to manage light.

How to Determine Optimal Distances for Subject Matter

1. Place your camera on a mount that can be adjusted vertically from approximately 1/3 to 2/3 the height of the room you plan to use for your shoot. When a table or

perhaps a projector stand is in use, we utilize a table-top tripod. Our favorite cameras have a standard ¼"-20 tripod mount, which allows us to use a variety of mounts.

2. In a flat room, the height of your camera should be adjusted based on the goals of the production. If you are shooting an educational event with a presenter and audience, the typical height is the midpoint between the height of the presenter's eyes and the average height of the participants' eyes. If you are trying to promote the speaker as an expert or superior to the group, you may wish to consider lowering the camera. If you wish to soften the speaker or increase the perceived superiority of the group, you may wish to increase the height of the camera slightly.

3. Every camera has some extent of lens distortion. 360-degree cameras use wide-angle lenses for video acquisition. The amount of lens distortion can be affected by the quality of the lenses and the number of lenses used on the camera. The distortion is often accentuated when the subject is very close to the lens or near the stich points. Turn on your camera now and put your face directly in front of the primary, forward-facing lens as you view the live preview on your connected preview device. You should see some distortion or aberration. Slowly back away from your camera until the distortion is alleviated. With most cameras, you will see noticeable improvement at about 18 inches from the lens and clear video anywhere from two to three feet from the lens. Take a measurement from the lens to the point where the distortion is alleviated.

4. Make sure you are still in the live preview mode. Continue to back away from the lens until one of two things happens: your image becomes distorted, or you (the subject of the video) become so small that the video does not have the desired level of detail. This is a very subjective distance and may vary from project to project. As a guideline, we use the distance at which we cannot clearly distinguish the difference in color between the

colored part (iris) and the white part (sclera) of the eye. Take a measurement from the lens to this point.

5. The two measurements you have are what we define as working boundaries for a shoot using this camera.

6. Now take the average of these two measurements, and this is what we consider the sweet spot for your important subjects. Remember we are working in a sphere. What we just did was define a linear measurement. You can use this distance in any direction from the lens. For example, if you were doing a virtual tour of a home and wanted to show a decorative ceiling, skylight, or fixture, you would need to make sure the distance from the lens to the desired subject is in the working boundaries and preferably close to the sweet spot.

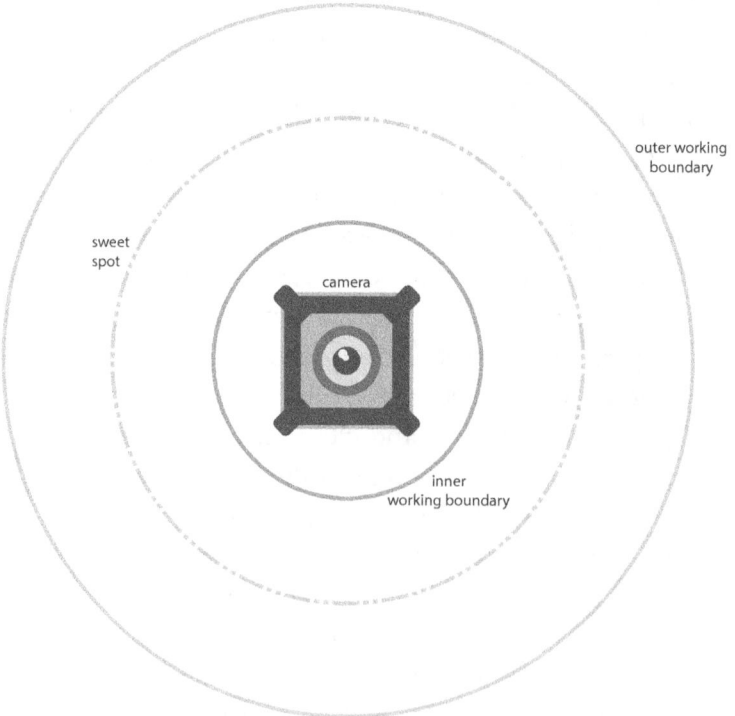

Figure 4.4

For proper camera placement, it's important you that understand your camera's working boundaries and sweet spot.

To Determine Stitch Points and Blind Spots

As we discussed early on, 360-degree video is a result of at least two images stitched together. Depending on the individual camera, videographers may experience a variety of visual issues because of this process. Depending on the lens distortion, the number of lenses, and the quality of the algorithms, you may see distortion, bands, or fuzzy video. Some cameras use waterproof housings and lens covers; with these, you may see an artifact or a distinct band. Conversely, depending on the lens configuration, some cameras also have blind spots. Stitch lines and blind spots are easy to find if you have a live view function, but it's a little tricky to find them if the camera just has record and playback. You will want to pay very close attention to how you complete the next few steps so you can replicate the procedure repeatedly.

1. Make sure your camera is on and in live preview mode.
2. Take two fingers and place them directly in front of the primary lens. You want to stay within an inch (a finger's width) of the camera's lens as you complete steps 3 and 4.
3. Slowly move your hand to the right and keep your eye on the preview device. Depending on your specific camera/lens configuration, you may see your fingers disappear anywhere from 60 degrees to 90 degrees from the center point of the lens. Remember, you are working in a sphere, so that blind spot will be circumferential like the working boundaries and sweet spot.
4. As soon as you find the blind spot, stop moving your hand and record the approximate angle.
5. Find your blind spot again with your fingers within an inch of the lens. Slowly move your fingers in a straight line away from the camera at the approximate angle where the blind spot was detected. As you get farther from the camera, the blind spot will dissolve into video footage. This is the stitch line.

6. Next repeat steps 3–5, except move your hand to the left instead of the right. You should be able to find the blind spot and corresponding stitch line.

7. Lastly, repeat this for each lens and note the overlapping regions of the blind spots. We'll discuss how to use these to your advantage in chapter 5.

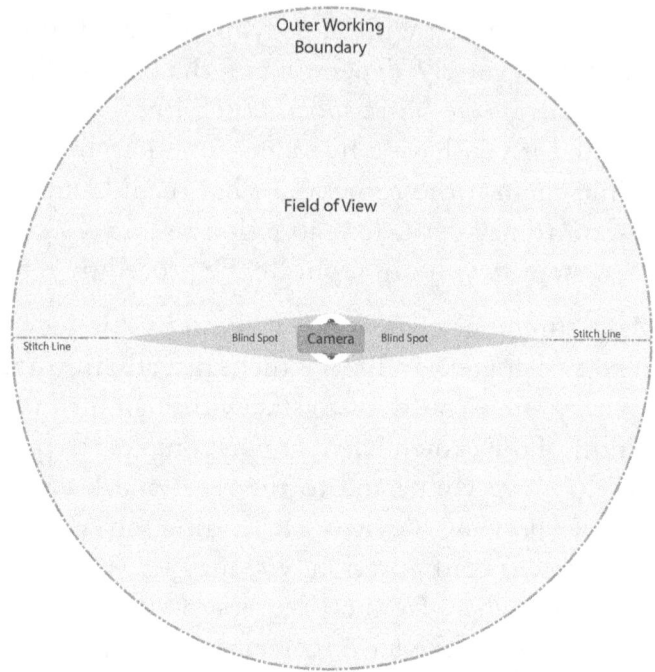

Figure 4.5
Experienced videographers will understand and use a camera's blind spots to their advantage.

If you are using a camera that does not have a live preview mode, the same exercise can be completed, but it's much more difficult. The process will require you to use similar movements as you record footage and then play back the footage as if you were using a live preview mode. You may need to repeat this process several times to obtain your stitch lines and identify your blind spots.

Videographers using either method may also find that using a paper grid may speed up the process. Your local office supply store will most likely carry large paper for a presentation easel. Purchasing a pad with lines or printed with a grid will allow you to quickly measure and mark your stitch lines and blind spots. Simply place the camera in the middle of the sheet and mark it up accordingly.

How to Check Lighting Settings

Understanding how different light sources will affect your footage is essential. Working with light within a sphere is different than traditional video. We'll discuss how to manage light in chapter 5. Pay attention to what works and what doesn't work so you can replicate these findings in the future. Here are a few steps you can take to understand how your camera will work with different light sources.

1. Turn on your camera and activate the live preview mode.
2. Start with normal ambient lighting for the room you selected for this exercise. Make gradual changes to the natural lighting sources by turning lights on or off, opening or closing blinds or shades, etc. Take note of what, if anything, happens to your feed on your preview device. Try to increase or decrease light to the point where your footage becomes noticeably different. Once you've made a significant change, access the camera settings menu and adjust the various settings available on your camera until you have adjusted the feed so that it is close to what you originally started with.
3. After completing step 2, change your camera settings back to where they were before this exercise and return the lighting to the normal ambient light configuration you started with.
4. Using a light box and reflector, bounce light throughout your field of view while monitoring your preview

device. Depending on the sensor capabilities and settings on your camera, you may create a significant change that will require you to make adjustments within your camera settings, or your camera may auto adjust. It's important that you know how your specific camera will be affected by this exercise.

5. Finally, turn on the LCD projector connected to your computer and bring up a white screen. (It could be a landing page for a search engine, a blank word process-ing file, etc.) Project the image on any wall in your room and position the center of the primary lens of the camera so it is centered on the image on the wall.

6. Evaluate if you can read the text or if it's simply a white image.

7. What happens if you simply rotate your camera lens left or right? Does the preview monitor show any change? Try moving the camera closer to or farther from the wall. What happens?

8. Change the color of the image. (You can make a new image, go to a different website, etc.)

9. How does your camera respond to the darker image? Rotate your camera left or right; is there any change with rotation? How does changing the distance from the lens to the wall affect the image?

How to Evaluate the Camera's Microphone

For this step (unlike the previous ones), we do not recommend using the live preview mode. The only way to truly evaluate the capabilities of your camera is to record some live footage and play it back. When you play back the footage, here are a few things for you to consider:

• Most productions will never use camera audio in a final production. A voice-over, sound track, or audio from a quality recording source is often inserted during the postproduction process. If your audio is relatively clear, it can be used to help synch your footage to another audio source if necessary.

- Some of the software playback options on the market do not allow you to make an accurate assessment. For example, one platform we evaluated while preparing to write this book generated a playback in which the audio was not synched with the video, but when the video was exported to a file, everything was fully synchronized.

- Just as video can be edited and tweaked in the postproduction process, audio may also be adjusted or edited. Like video, every effort should be made to acquire a quality audio recording, but there are other options when things go bad. (Refer to chapter 9.)

Exercise 2: Outdoor Shoot

Once you have mastered your camera indoors, it's time to move outside. Over the years, we have worked on hundreds if not thousands of outdoor projects. Both experienced and new videographers may find certain outdoor shoots challenging. Although the operation of your camera and the information and measurements you have acquired in exercise 1 will remain constant, taking your production outdoors immediately places you in a dynamic environment.

At the conclusion of this exercise, you should be able to determine an optimal camera position for an outdoor shoot.

This exercise works well in just about any outdoor environment. We recommend you repeat this exercise in many different locations so you can truly understand not only how your camera will capture different subjects in different settings but also how you can change the viewer's experience by simply changing the height and rotation of the camera.

In addition to your camera (and the ability to preview it), you will need an adjustable mount that will allow you to change the vertical height of the camera. We also suggest that you again use a tape measure and record your findings for future use. A reflector and standard golf umbrella will also allow you to experiment further with outdoor lighting.

Before you take your camera outdoors, there are many things to consider. We cover these in detail in chapter 5. For the experiment, it's important that you are mindful of how changing weather might affect you and your camera.

1. Get outside.
2. Go to a park, parking lot, wooded environment, or waterfront. Each will offer you a different opportunity to place and adjust your camera.
3. Affix your camera to your vertically adjustable mount and put it in live preview mode.
4. Define your primary subject for the video.
5. Using the measurements you acquired in exercise 1, place your camera so the subject is inside the working boundaries, with your subject as close to your camera's sweet spot as possible.
6. Using your preview device, what do you see? Look up, look down, then look all around the sphere. Does everything look accurate? Do you notice anything that needs to come out of the shoot? If you are shooting a promotion for a high-end restaurant, you probably don't want a garbage can in the field of view. It's almost always possible to remove something or mask it in postproduction; however, it's much quicker and more economical to take the time to move the object at the time of shooting.
7. Keep looking. How does the lighting look? Are there any shadows? Is anything washed out? Do you need to add light? If so, where do you add it and at what angle to make it look natural?
8. Take out your reflector. Can you bounce or mask the light source to change your video acquisition?

9. Put your umbrella over the camera. What happens? If you look up in live preview mode on your device, you should see the umbrella. Can you see a shadow? Put the umbrella away.

By now you should be getting comfortable with a few things. You should be good at setting up your camera. You should be familiar with your camera settings and using the live preview mode. You should also be comfortable with changing your camera's adjustable settings.

Because of the vast number of variables in the setting within a camera and the unique environments you will be shooting in, it is impossible to provide you with specific configurations or parameters for your individual shoots. What we can offer is the suggestion that you should shoot everything and anything as often as you can. The more you practice with your camera, the better results you will have; if you do this to pay your bills or have a desire to do so, the more business you will generate.

Exercise 3: Water

You should now be comfortable with your camera and the various positioning and lighting challenges both indoors and outdoors. Let's introduce you to water.

At the conclusion of this exercise you should be able to discuss the additional considerations and required actions when filming in water, the capabilities of your camera, and how to mount a camera in water.

This exercise is designed to incorporate water into your workflow. You can do this in a sink, a fish tank, or a bath tub. We've experimented in fresh water with kayaks, jet skis, and a variety of boats. The principles are the same, but there are a few differences in the mounting of the camera.

Before we go any further, in this case it's *essential* that camera operators understand the capabilities of their individual cameras.

While not knowing all the frame rate capabilities may not get you into trouble, not knowing if your camera is waterproof before it goes in the water could be an expensive mistake. Make sure your specific camera is truly *waterproof!* If it says "water resistant," do not complete this exercise or use the camera near water. Quality camera manufacturers will not only tell you if their products are waterproof but will also provide you ratings on the seals, which are usually conveyed by how deep you can go and/or for how long the cameras can be immersed in water. Some cameras require specific lens protectors or the use of a waterproof enclosure. Now is the time to figure this out and make sure your investment is protected.

If your specific camera is not waterproof but you are interested in shooting in the water, an after-market enclosure may be a viable solution, for these enclosures are often less expensive than a new camera platform.

The live preview mode will not always be a function you can use in water filming. Years ago, my company purchased a traditional sport video platform. In addition to the standard camera package, we bought additional floats, waterproof enclosures, and a dedicated tablet to use for our live preview mode. Shortly after that, we were retained for a project to capture footage at an exotic fish distributor to develop a video series that would promote their offerings. We arrived early at the shoot. Our gear was prepared, and we were ready to go. We recognized that we had been awarded this project after two companies had been unsuccessful in providing the client with the video they were looking for. As we were getting ready to shoot the first tank of clarion angelfish, with a retail cost of $2500.00 per fish, we were answering questions and addressing concerns about the equipment we would be putting in the tank. The owner wanted to confirm that our gear would not harm the fish or create any disturbance in the water quality. As part of our assurance, we suggested the owner hold the tablet and monitor camera in the video preview mode. This was normal and something we had done with clients many times. We completed our final checks and ensured the enclosure around the camera was secure. We started the recording. Everything was looking great. Then we put the camera in the water, and the screen went black. The owner screamed and dropped the tablet. The videographer quickly removed the camera from the tank, and the image returned to the tablet.

What we learned the hard way was that the Wi-Fi signal from the camera was blocked by the water. Duh! Radio signals are disrupted by water. Wi-Fi and Bluetooth signals will be degraded by even a few inches of water. We were tech guys that lived in a tech environment. We had used the Wi-Fi features of the camera and tablet combination many times, but never under water. We had read the manuals on the cameras and the app. There was no mention of this limitation.

After a rocky start, we did capture the footage in several tanks and deliver a product that the company used to promote their business for several years.

If your live mode preview depends on Wi-Fi or Bluetooth, you can't use it under water. This means that if you are shooting underwater, you need to plan time to mount your camera, shoot some footage, remove the camera, review the footage, and replace the camera. For this type of shoot, it's not acceptable to simply record your footage without any preview.

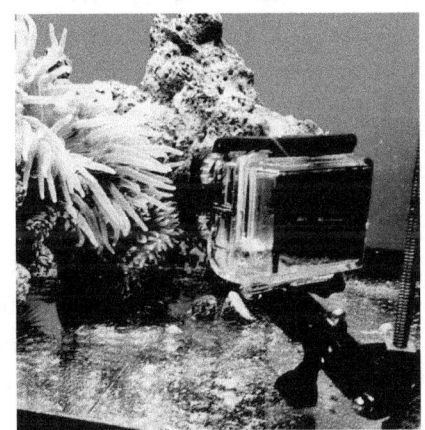

Figure 4.6
Wi-Fi will not transmit through water.

Over the years we have found two exceptions in which live preview mode may be used with water.

1. A few traditional camera platforms support a cabled live preview monitor. While we are not aware of any 360-degree platform on the market that currently has this feature in a waterproof setup, it's likely that one will eventually be introduced.
2. On occasion videographers may wish to split the sphere and show part of it above the waterline. In this case, depending on how the camera is mounted, the transmitter may be positioned above water and still enable Wi-Fi or Bluetooth to transmit preview data.

After you have confirmed that your camera is waterproof, you need to select a mount. While these next few points may seem obvious, I believe they need to be made. Unless you purchase a specific float for your camera or enclosure, your camera will sink. If it sinks, most likely it will be reclassified in your accounting system from an asset to a loss, because you will probably not get it back. Take the time to choose the proper mount. We cover mounts extensively in chapter 5; however, here are a few specifics for mounting in water.

- If you are not able to work with a dry surface, such as the hull of a boat above the water line, you should use a mount that allows you to secure the camera mechanically. Spring clamps, handle bar mounts, belt mounts, and even tripod mounts can sometimes be modified to provide a mechanical or friction mount.
- If you have the ability to work with a dry surface, many quality tape mounts are available. There are also many after-market knock-offs. Do not try to save a few dollars

by using a knock-off tape mount. They are not all the same, and in water there is a noticeable difference. If you are going to have a mount fail, the last place you want it to fail is in the water.

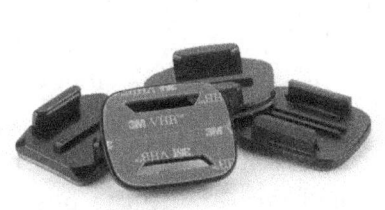

Figure 4.7
Tape mounts are not the place to save a few dollars on your production budget.

• No matter what mount you select, have a back-up for it. We use a simple nylon cord as a tether with all our water mounts. (On occasion we tether other mounts too, but without fail all water mounts are tethered.) There are two simple points to remember: the cord must be able to support the force created by the camera should the primary mount fail as well as the weight of the camera, and the tether should be affixed to a different area than the point where the primary mount is affixed.

Once your camera is mounted, record and review some test footage. Camera position and lighting settings are the biggest challenges of shooting in the water. You may need to adjust the position of your mount or the camera settings. Some cameras have specific settings for shooting under water. These settings usually create results similar to those you would get by manually adjusting your lighting controls within the manual settings. You may need to repeat this process several times before you are ready to record quality video.

Exercise 4: Extended Field of View

360-degree video works well in many environments. One area where you will see exceptional results in comparison to traditional video is capturing multiple subjects in an extended field of view. Traditional video uses a single lens, which forces videographers to adjust a limited field of view to capture multiple subjects. The videographer may pan, zoom in or out, or a use a defined cut during the postproduction process to show multiple subjects in a smaller field of view. With 360-degree video, working in a sphere allows the videographer to capture multiple subjects without the need for these other techniques.

At the conclusion of this exercise, you should be able to determine an optimal camera position for multiple subjects in an extended field of view.

This exercise works well in a large room with a variety of large stationary objects at a variety of heights. We have completed this exercise in an elevated lecture hall, the bleachers at a sporting event, at a warehouse, and on a trail alongside a waterfall. Each of these settings had different objects at different places around the sphere as well as at different heights.

In addition to your camera and the ability to preview it, you will need an adjustable mount that will allow you to change the vertical height and (if possible) the tilt of the camera. By now you should be an expert with manipulating the other settings within your camera.

Here's how you set it up:

1. Start by placing your camera in the center of the sphere so all of your desired objects are within the working boundaries of the camera and as close to the sweet spot as possible.

2. Should your subjects not be equidistant from your lens, you may wish to prioritize your subjects and adjust your camera's position so the subject with the highest priority is in the camera's sweet spot.

3. Your camera should be placed in the sphere level with the ground. If you do not wish to emphasize one subject over another, the starting height should be approximately one-half the distance from the lowest to the highest point of the desired subject.

4. Should you wish to emphasize one subject over others, that subject should be closer to the lens than the others and ideally in the camera's sweet spot.

As you have gone through these steps, have you moved your camera up and then back down? Or maybe from left to right and then left again? This is not uncommon. It's part of the challenge of working in the sphere. The four steps above are designed to help with an initial camera placement.

Once you have completed these steps, additional adjustments may be required to fine-tune your imaging. In addition to moving the position of the camera or the height of the camera, you can also adjust the tilt. Moving the tilt may accomplish several things when fine-tuning your shot.

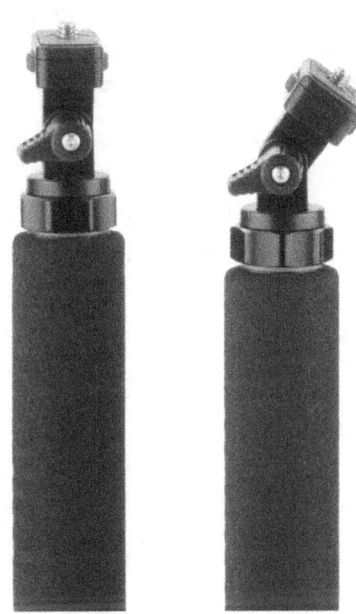

Figure 4.8
Tilting your camera may help you fine-tune your shot.

Consider this example. You've been hired to shoot promotional video for a band at a live performance in a nightclub. The client wants the focus of the footage to be on the lead singer of the band and her ability to get the patrons to dance.

Before the performance, you visited the club and established that the lighting is sufficient. The band and dance floor are within the working boundaries of the camera, and if the camera is attached to the front of the stage in front of the lead singer, both the patrons and lead singer are relatively close to the camera's sweet spot. Because the camera is closer to the lead singer than the dance patrons, the footage will convey a subliminal emphasis on the lead singer.

So far we have discussed moving the camera left to right, front to back, or up and down. Using this example, if we move the camera higher, the lead singer becomes more pronounced but the patrons fall back and get distorted. Tilting the camera raises one side of the camera and lowers the other, which changes the position of the subjects in the sphere but will minimize any distortion or unintentional prioritizing of subjects by changing their distance from the lens.

Tilting a camera may also produce other subtle changes in your shot, such as changes in lighting or the camera's ability to record detail in shadows within the sphere, improved quality when capturing video or slides on a projection screen, and reduced glare on glass from natural sunlight.

Chapter 5

Planning the Shoot

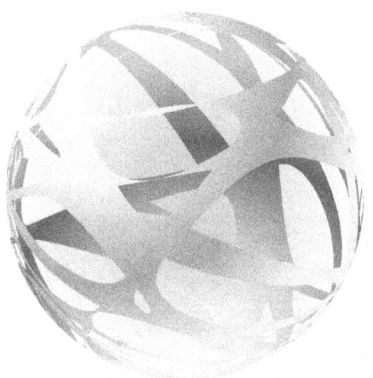

There are many aspects to a successful shoot. In chapter 6 we will take you through the process of a shoot from start to finish. This chapter covers items that you should consider before you begin a shoot.

Camera Mounts

Planning a shoot starts with the basic question: "Where should the camera be mounted?" We covered the basics about camera positioning in a previous chapter; once you have figured out your positioning, it's time to mount the camera.

For a period of time my video division used the tag line, "We get the shots our competitors have not even considered." We were able to promote this and continue to deliver quality video products to clients for many different reasons, but one key thing that continues to differentiate us from our competitors is the way we mount and move cameras. We've had cameras on golf carts, cars, trucks, motorcycles, boats, jet skis, ultralight aircraft, and helicopters. We've had them under water to document the movement of sea turtles, several hundred feet above the forest canopy along a zip line, and on top of radio towers to capture bird migration. They have been in shark tanks, robotic ware-houses, and operating rooms. We've used cranes, jibs, tracking dollies, sliders, skates, helmet mounts, and chest mounts, as well as various combinations of all these mounts. Our best shoots and most successful productions always have at least one shot that creates the "Wow!" effect.

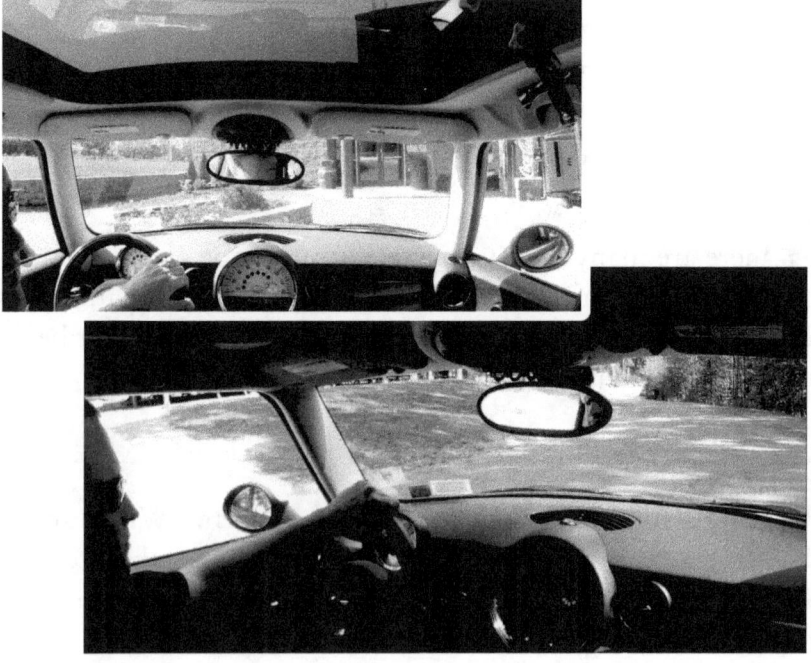

Figure 5.1
An example of a creative mount and a screen shot of the image acquired.

A successful camera mount always has four basic principles. Consider the following example. You have been hired to shoot a commercial featuring a new car. The goal is to showcase the new body style that makes it different from last year's model. Here is what your mount should include:

1. **It should provide an interesting and appropriate perspective that aligns with the goals of the production.** Many videographers might set up a camera on a tripod and simply pan through the shot. Consider something unique: a slider with a crane or a suction cup mount that captures the curves as the car is moving along a scenic landscape.

2. **It should help the viewer focus appropriately on the desired subject matter.** If you want the viewer to focus on the new body style of a car, your camera should be positioned to capture the curves of the sheet metal. Many times, you might be tempted to capture footage you may feel is more interesting. In this example, this might include shots of the engine compartment, the passenger cabin, or even the undercarriage. If these will not overshadow your goals—get them! Call it the B-roll. You may use the footage for this project, and having it in the can might be of benefit later on. But always, always make sure your primary mount/shoot is focused on the goal.

3. **It should help the viewer have the desired response.** If you want to sell cars, you need to evoke a favorable response from the viewer. In this case, you may be showing quality workmanship and cool colors, or perhaps even establishing emotional response parallels with a current event in the news, music or fashion trend, or popular opinion. Simply changing the height of the camera from the focal point may evoke a feeling of inferiority or superiority. Having a mount that can sustain high speeds and a camera at a high frame rate that is slowed down may draw a viewer in. Put yourself behind the eyes of the viewer. What will make the results interesting?

4. **Your mount must be absolutely safe and secure.** This is the most important principle. Our team has a very diverse background, and we have been involved in many different situations over the years. We've seen the "good, bad, and

ugly" with not only video production, but also life in general. If something can go wrong, it will eventually happen. Don't let a bad mount become a career-limiting move. Numerous mounts for cameras are available today; while many are compatible with a variety of cameras, not every mount is appropriate for every situation. If you are shooting the side of the car with the camera stationary and the car moving, you could choose almost any mount, but if the camera is to be mounted to the car, you may wish to use a clamping device, a suction cup system, or perhaps even a specialty adhesive tape mount. Depending on the anticipated speed of the car, you may wish to modify your selection and add a tether. A camera that becomes a projectile can turn a simple shoot into a catastrophe. Take your time, double- or triple-check your setup, and consider all your options before calling for action.

Whether you are mounting traditional video cameras or cameras capable of 360-degree video, the rules are the same. You must consider distance, lighting, and secure fastening of cameras.

Our discussion of mounts is divided into three categories: tripods, specialty mounts, and dynamic mounts.

Tripods. You do not need an expensive tripod for 360-degree video. Since you are recording the motion in a sphere versus creating motion with a camera, you have no need for an expensive fluid head. In fact, a small lightweight or pistol-grip head is preferred over a traditional pivoting head.

Most cameras are equipped with a standard ¼"-20 tripod mount. This allows the camera to be mounted securely to the top of a tripod. Although most videographers have mastered the use of a tripod for traditional video work, using a tripod for 360-degree video may present an additional challenge. Again, remember that you are working within a sphere. If your primary subject is straight out from the lens near the horizon, a straight or level radius on the horizon, or even looking up, a tripod may make sense and be a viable choice. But what happens when the viewer wants to look down? In addition to seeing the desired

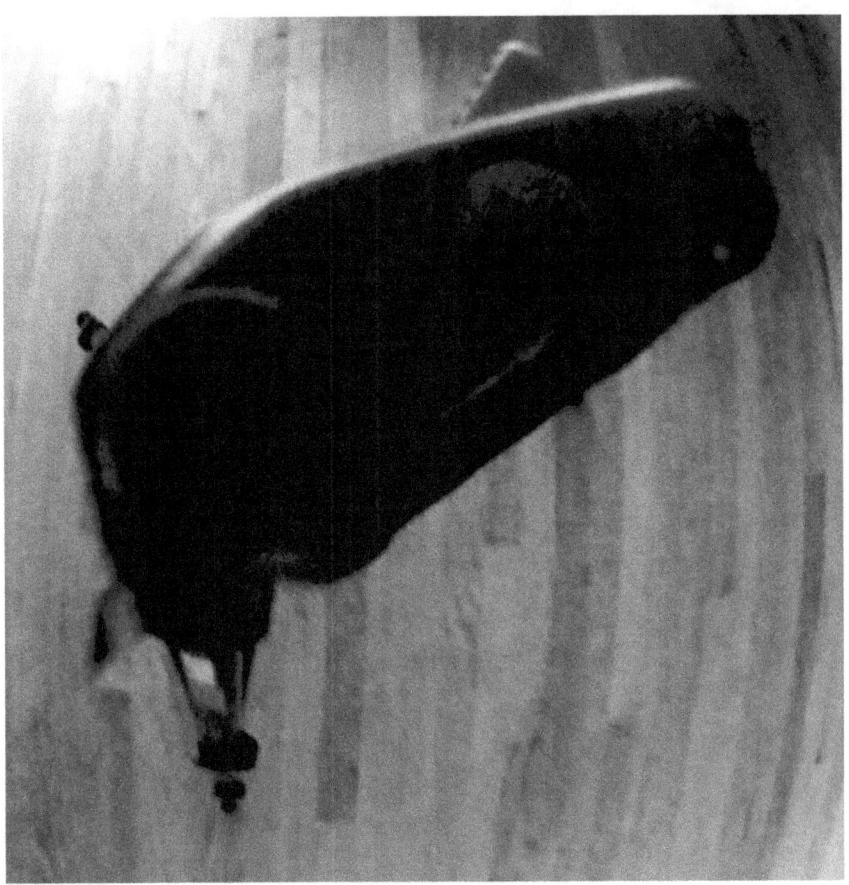

Figure 5.2
Tripod legs in the field of view within a sphere.

perspective or subject, the tripod legs will be within the field of view. Depending on your shot, this may or may not be an issue.

Consider what you're shooting. If you are shooting an air-show, the field of view and focus of attention will most likely be away from you on or above the horizon, where the planes are taking off or landing on the runway or performing in the sky. Users will have little if any interest in looking straight down at your feet or the grassy field and seeing the tripod legs within their field of view. We are not suggesting that no one will ever look down or that it is acceptable to have a view in the lower part of the sphere that is cluttered with or obstructed by equipment, but your subject matter is something to consider. If it makes you feel better, we are not alone in our thinking, as some cameras on the market only shoot a 240–270-degree sphere.

If a tripod is your only option, there are two ways to deal with this. First, you may just leave the video as it is; should the viewers look low, they will see the tripod legs as they normally would appear in the sphere. The second option is to edit the tripod legs in the initial footage during postproduction (which will be addressed further in chapter 7).

Specialty Mounts. There are several different amounts that fall into this classification. Anything that is not a tripod is a specialty mount. (Dynamic mounts have one additional attribute that we will cover in the next section.)

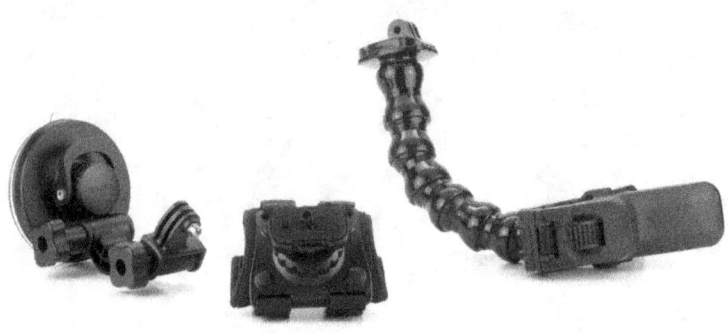

Figure 5.3
A wide assortment of specialty mounts provides many options for mounting a camera.

Dynamic mounts include body mounts, clamps, belts, specialty brackets, tabletop camera stands, and even tape mounts. These mounts are designed to hold the camera in a fixed position during the acquisition of video. When selecting a specific specialty mount, you should base your choice on the answers to these three questions:

- Is it capable of securely attaching the camera to the desired object?
- What parts of the mount will be visible during video acquisition?
- How easy (or hard) is it to set up?

On a typical production, we may carry up to ten or fifteen different types of mounts. Once we determine the ideal position of the camera, we select an appropriate mount. Over the years we have found that simple tape mounts, a clamp-on mount designed for a handlebar or roll bar (we carry both sizes), and a spring clamp with a flexible neck are great choices. These work well for traditional video as well as 360-degree video.

After many shoots and spending hundreds of dollars on a variety of specialty mounts, the mount that gets the most use in our productions (including 360-degree video productions) is something many videographers already have in their studio—a light stand. Most light stands have a ¼"-20 mount on them. (It may be covered or capped; see the photo below.)

A light stand has several features that give them advantages over other mounts. They are stable, yet offer a low-profile leg system in comparison to a tripod. They offer a wide vertical range of adjustment. (Our stands can be adjusted from three to nine feet in height.) They are lightweight and easy to carry, and can be set up or broken down in a matter of seconds. They are inexpensive. Some are also made of materials that allow them to be used inclement conditions. (Ours are aluminum.)

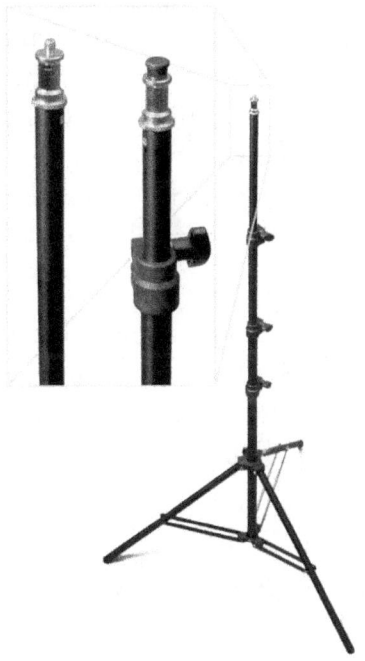

Dynamic Mounts. All dynamic mounts are specialty mounts, but not all specialty mounts are dynamic mounts. The one factor that separates the two categories is the ability

Figure 5.4

An example of a basic light stand. Almost all have a 1/4"-20 mount included. This shows one mount exposed and another covered by a thread cap.

for the mount to hold on or remain attached when an external force is applied.

As we mentioned previously, not all tape mounts are created equal. Some use a name-brand adhesive tape that has undergone specific testing and is appropriately rated for use in special situations. Other tape mounts have a similar connection to the camera, but the tape may be a generic double-sided adhesive or a foam-core tape that is also sold to hold posters to a wall. Other examples of dynamic mounts include body mounts, custom brackets, and suction cups. Again, we recommend that all dynamic mounts should use a tether whenever possible.

Lighting

As previously mentioned, 360-degree video cameras rely on wide-angle lenses with a large field of view. One of the limitations of these cameras is their ability to adjust to various light sources and differing levels of intensity and quality within the field of view.

When you are shooting traditional video, lighting adjustments are relatively easy to make. Typically, your options include:

- Repositioning the camera
- Strategically adding direct light and/or fill light
- Diverting light with a reflector or screen
- Changing an exposure or ISO setting within the camera

Because 360-degree video makes everything visible within the sphere, it's not as easy as adding a light box or handing an intern a reflector to bounce some light towards the primary subject. A 360-degree video camera requires some creative solutions.

- You might not be able to simply move the camera, but you may be able to move an item or introduce a similar item closer to the lens.

- If you have a window that is in the shot, you may be able to make adjustments by either using an external light source or reflector to bounce more light through the window or placing a tarp or screen to filter out some of the light.

- If you have incandescent floor or table lights within the shot, you may be able to use a dimmer box to turn down or reduce the light. You may also need to add additional light sources within the shot, such as additional floor or table lamps that fit the scene. We recently did a shot in a repair shop in which we used the lights on a vehicle to add more fill light. Be creative.

- One of the biggest challenges is working within an environment that has a projection screen showing either a static slide presentation or a video. This often creates a significant abnormality within the sphere. The easiest way to address this situation is to plan ahead and minimize the impact. If you can't eliminate the screen from your shot, have the presenter prepare a slide deck that uses a darker background of blue, green, or burgundy with white letters for an optimal contrast. Work with the presenter to select a video that is not made up of light background or bright video clips. If you can't get the presentation changed, two other options have been successful for us. One is to increase the ambient light in the rest of the sphere to minimize the contrast with the presentation, and the other is to try tilting the camera. Although it is the job of the videographer to capture the events in the best way possible, this is something that is not only difficult, but often despite what you try, it does not make for great video. We believe it is essential to identify this early on in the project and bring it to the attention of the client you are working for. No matter what they tell you, clients do not like surprises, and pointing this out ahead of time will make you a champion if you can pull it off or soften the blow should you have difficulties.

Staff, Bystanders, and Production Boundaries

By now you should clearly understand that in the sphere, the 360-degree video camera will capture *everything* and *everyone*. For this reason, you should minimize your crew and control access of bystanders to the shoot. When the word gets out that you are shooting 360-degree video or virtual reality, there will be a lot of excitement. Many will want to watch and see what the magic is all about. Unfortunately, when they are watching, there are very few places to hide.

Most adults will not fit within the camera's blind spot that you identified in exercise 1 in the previous chapter. You really only have two choices;

- Using your live preview mode, establish production boundaries. Measure out the working boundaries you determined in chapter 4 and then add about twenty feet. Have your bystanders assemble at that point and verify they cannot be seen.
- Hide your staff or bystanders behind objects that create blind spots within the sphere, as figure 5.5 illustrates.

There will be times when you want to feature bystanders or staff members within the sphere. Then it's important that they be kept within the working boundaries. Depending on the specific situation, you may need to provide marked areas or production boundaries for participants to remain in. Whenever possible, use items that are already within the shoot to establish these boundaries. If you need specific markings on the floor, clear packing tape is our method of choice.

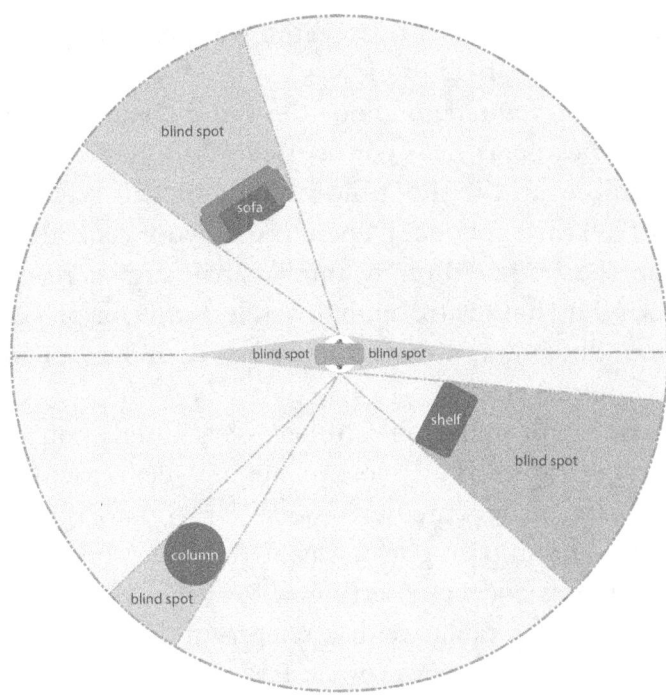

Figure 5.5
Blind spots may be used to hide other cameras, gear, or crew members.

Mics, Mixing Boards, Monitors, and Teleprompters

Again, the 360-degree video camera "sees it all." Think about all the equipment that gets used on a typical video production. We've already talked about lighting, but what else? On our last shoot, if you looked quickly, it was like making a visit to a camera shop or distributor. We had boom microphones, audio cables, sound boards, audio recorders, camera monitors, a producer monitor, and a teleprompter in use. The camera was on a 16-foot jib that sat on a medium-duty tripod within a tracking dolly, which was on eight feet of track. Behind the lens we had a proofing laptop, external drives, and scribe. We also had a stack of travel bags, high-impact plastic protective cases, and shipping containers. It was intense. Everything was positioned out of the traditional camera's field of view.

What if we wanted to shoot the same scene using a 360-degree video camera? Let's consider everything piece by piece.

For audio gear, the shotgun microphone gets replaced with a lavalier (lav) microphone. Depending on your capabilities, you may drop the output directly into a small digital recorder or perhaps send its output through a wireless unit for pickup. The cables, soundboard, and recorder need to be moved outside the working boundary.

Camera monitors become unnecessary unless your camera has streaming capabilities; most of the 360-degree video cameras offer only a live preview mode for you to set up your shot. The teleprompter goes away. Depending on your project requirements and your setting within the sphere, you may be able to utilize a tablet with a teleprompting app, cue cards hidden within a blind spot, or flash cards that are referred to between shots.

We've already discussed camera locations. In the next chapter, we will discuss moving shots with dynamic mounts. In this example, we would simply replace the traditional camera on the jib, tripod, tracking dolly, and track with a 360-degree video camera on a light stand.

We would then move all our proofing gear, scribe, and shipping bags, boxes, and containers well outside of the working boundaries. Working within a sphere is not difficult; it just requires a different mindset.

Power

Like audio cables, it is not desirable to have power cables for battery packs, corded connections, or generators within a sphere when you are working. As we discussed in chapter 4, understanding your battery life is important. Make sure you have enough battery coverage for the shoot, plus a spare battery.

Outdoor Shoots

Planning for an outdoor shoot requires a similar approach to shooting indoors; however, there are some additional consider-ations you should address before you take your camera outdoors:

1. Is your camera able to work in the temperature range you wish to expose it to?
2. What is the weather now? What is it going to be like in an hour? Have you checked the weather radar? Do you have the ability to monitor weather developments throughout the shoot?
3. If it's raining or snowing, is your camera waterproof?
4. Where are you going?
   a. Are you dressed appropriately? Staying dry, warm, or cool and having the right footwear will let you stay focused and ultimately allow you to produce a better product.
   b. If you are going to the beach, can the camera tol-erate blowing sand, salt water, etc.? Even if it can, do you want to expose your gear to these elements? (Unless special circumstances are present, we often turn down work on a beach for these reasons.)
5. Do you need special equipment or mounts?
6. Do you have the power source(s) you need, such as bat-teries or a generator?
7. What are you going to shoot? Shooting people is differ-ent than shooting a static object, which is different than shooting a moving object.

Data Management

One of the best things about being a videographer today is work-ing in the digital world. We will talk about this more when we discuss postproduction in chapter 7 and distribution in chapter 8. Having a data management plan is essential for amateur and professional videographers alike. Although individual settings and file formats can affect the file size, generally speaking, traditional video uses significantly more storage space than still photography, and 360-degree video uses more storage space than traditional video.

Videographers must have at least a basic understanding of file transfer and storage on the computer platform of their choice as well as have a plan to address the following issues:

- **Transfer times.** How do you plan on transferring the data, and how long will the transfer take? Do you have the right equipment, such as a card reader? Are you saving the data to a local drive, a network, or the cloud? The use of Wi-Fi or Bluetooth for transfers will be significantly slower than a wired connection.

- **Storage of data.** Before you shoot, you should estimate how much data your camera(s) will generate on the shoot and make sure you have the ability to store and manipulate the data appropriately. As you develop this estimate, don't forget about additional data sources such as audio files, client-supplied graphics, or voice-overs. A good guideline is making sure you can manage one-and-a-half times the amount of data you think your shoot will generate.

- **Redundancy.** How many copies of the data are you making? Are you keeping the data on the cards intact until the project completes postproduction, or are you simply transferring the data off of them and cycling them back in the shoot or putting the cards back on the production vehicle? Our years of experience have taught us to retain at least two copies of every file for every project. In some cases, we have three or four copies of the project files. Why? Because things go wrong.

- **Disc drives and failures.** When my wife and I were married, one of our wedding gifts was a digital camera. As early adopters of that technology, we abandoned our

35 mm SLR cameras and started shooting digital. We took it on our honeymoon, shot the many "firsts" of a new marriage (such as our first house and our first puppy), and used it for all the typical family events and functions. One day my wife phoned me to tell me that our photos were gone. They were no longer on the hard drive and, yes, we had no backup. Today companies specialize in hard drive recovery. While they are extremely expensive and may not recover anything, they offer a chance for recovery, but back then these folks did not exist. After days of searching for assistance, reaching out to different vendors, and trying several do-it-yourself recovery programs, we wrote off the drive as a loss. (In hindsight, I should have simply put it in a box and had it restored for a future anniversary gift. Who would have known that new technology would have offered me a second chance decades later?!)

During this life lesson, I spoke with several customer service representatives and technical support teams from the computer manufacturer and the hard drive manufacturer, as well as third-party software vendors. One of the engineers made a statement that has stayed with me over the years. "You should have a backup plan for when your hard drive fails, not if it fails. All hard drives fail." Over the years I've used floppy discs, tapes, CDs, DVDs, and external hard drives for backups and redundancy. Today our workstations are backed up to a server, our client files are stored in a protected array, and all this data is backed up to two different locations. We are prepared and protected. If you cannot say you are protected with certainty, you need a better plan.

One last thing to note. There are many hard drives on the market and many are becoming specialized for specific functions such as workstation applications, file servers, or multimedia/gaming systems. The newest hard drives are solid state drives (SSD). Unlike traditional hard drives, which usually have one or more physical discs that spin and a moving arm to read the data, SSDs have no moving parts. In simple terms, they are like a data card, because they store data on a flash memory device. They are typically quicker than traditional drives, do not generate heat, and last longer. But make no mistake, these drives also fail, and the data on them should be included in your redundancy plan. While we have yet to have one of our SSDs fail, we have been told they fail with no notice. With a traditional drive, you might start to hear a motor whine, a bearing grind; such noises are usually an indication of an upcoming failure. With SSDs, they just stop working.

- **Mobility.** Does your data plan account for large projects? It's not uncommon for us to be on a shoot for an extended period of time. We may be shooting locally and retuning to our studio each night, on the road chasing helicopters, or perhaps shooting a project we had to fly to. Depending on the length of time we are away and whether we have one of our vehicles or have had our gear flown out, we adapt our plan for the individual project. The amount of data, our available devices, and the transfer speeds all play in to our decision. We may simply back up the data to a laptop or tablet and retain the cards from the cameras. In some instances, we may duplicate the cards to multiple portable external hard drives. For some projects we may load hard drives during the day and then use FTP to transfer the data to our server later in the evening and let it run all night.

In today's world, there is no excuse for lost data. Do not jeopardize your client's project and your relationship with your client by not having a solid data plan.

Chapter 6
The Effective Shoot

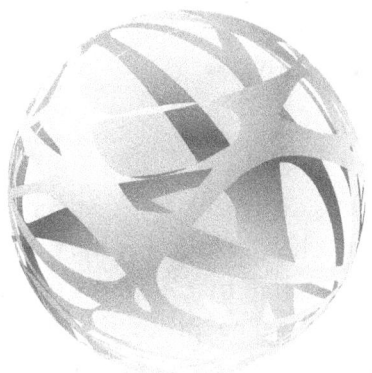

In this chapter, we will take you through the typical process we use to conduct an effective video shoot. While some of the information within this chapter is specific to 360-degree video production, a significant portion is applicable for traditional video production.

The intent of this chapter is to provide information for the amateur and professional alike. Depending on your individual experience and skill level, you may find some of this information very basic, a good review, or a completely new way of thinking. If you are a new videographer or someone starting a new venture, this chapter may save you hours of work or prevent you from making some missteps along the way. Looking back, had we had a resource such as this in our early years, our first few years of operation would have been less stressful and more profitable.

The information presented in this chapter is designed not only to provide you with some specific advice or direction but also to present you with additional questions you should consider along the way. While this insight reflects the experience we have gathered over many years, it is important to understand that no two studios, two projects, or two videographers function exactly the same way. It is important that you use this chapter as a guideline and adapt its advice to your specific situation, project, or studio workflow.

Proposal and Contracting

If you are completing a project for any type of compensation, an effective video production should start with a solid proposal and contract. Today we live in a litigious society. It is important that any business venture be protected by a solid production contract. More importantly, a good contract will not only protect you but also define the project and deliverables. It should address other important aspects such as billing terms and conditions, copyright and inclusion of copyright material, ownership of the final product, and any guarantees or warranties.

At our studio, we have found it more effective to use a general production contract that is relatively generic and applicable to all clients we work with. This document defines our engagement, the use of works protected by copyright, billing, and other general business matters. This contract has an auto-renewal clause that basically keeps it in effect until one party indicates they want to end the relationship. We use a second document that specifically defines an individual project. This model has allowed us to be extremely flexible with our terms, operations, and other defined items like ownership and rights associated with specific projects. By executing a long-term contract versus simply a per-project contract, we are demonstrating our desire for a long-term relationship early on. We have found that model also sets the tone of us becoming a partner, not a single project vendor and has led to more repeat projects with clients than we can count.

We then prepare a written plan that is specific to an individual project. It is presented to the client in the form of a proposal and requires their signature and approval before we move forward with their project.

If you are working on a project for enjoyment or don't have a specific client, you will find a plan helpful. A solid plan will force you to work effectively and ensure that your

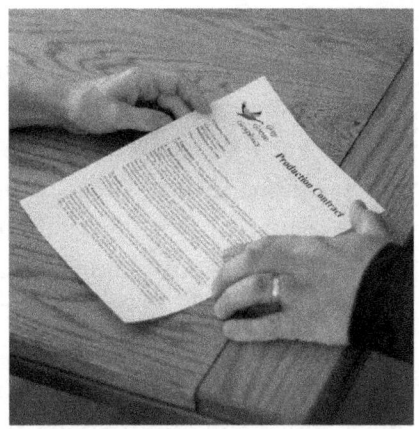

Figure 6.1
An effective video production should start with a solid proposal and contract.

production will be completed efficiently. Should you evolve into the role of a paid professional, you will be ahead of the curve with the management of your projects.

The project plan accomplishes three specific goals.

It defines the project. Within this proposal, we define the obvious basics such as the overall project goal, specifics relating to the dates and/or times that shoots will occur, the location(s), those participating, and the desired deliverable.

It defines the responsibilities of those involved in the project. We try to identify to the greatest extent as possible who is responsible for the different parts of the project or bringing the different assets to the project. For example, who is responsible for things such as arranging the location, bringing the talent, supplying an audio track or voice-over, or creating overlay graphics? Depending on the type or size of the project, this list may be concise or may be very extensive.

It defines the budget. Depending on the client, our existing relationship with the client, or historical information about the client or about a similar production, the budgets we present are compiled using one of the following models.

Flat Fee. This model presents the client with a single fee for all services to be provided. It encompasses not only our production labor but also any additional expenses required to complete the project from start to finish. These expenses might include travel, talent, location fees, graphic design, soundtracks, voice-over expenses, or other production expenses.

Although many clients prefer the flat fee model because it gives them a defined price upfront, it can be very dangerous to the inexperienced videographer and potentially more expensive to the client. If you own your own home, you can relate to this analogy. Remember the first time you took on any simple repair. What first seemed like a simple project probably resulted in multiple trips to the hardware store or home center. It most likely took you much more time than anticipated, and by the end of the project, it may have cost you more than your original estimate. You probably would have made a different estimate for your required time and materials had you done it again. This is very similar to the flat fee model for an inexperienced videographer. Without a lot of estimates and project history, inexperienced videographers may not realize what additional time or expenses are associated with a specific project and may soon find themselves at an economic disadvantage, perhaps even losing money. On the other hand, the experienced videographer understands that there is a risk with certain types of activities and will inflate a budget to cover these possibilities. The flat fee video production model will almost always cost the consumer more than other models if the videographer has experience.

Time and Materials. This has become our model of choice for many reasons. First, it provides the client with fair and consistent billing for the actual services and expenses that are incurred on their project. It gives clients an incentive to be prepared and ready for the shoot, for the more efficient they are, the more efficient and effective the production crew can be. This translates to a less expensive overall production. This model also covers the production team for any change orders or delays on the client's behalf.

From our experience, existing clients or those we have a history with find this to be a very desirable model. New clients may question this somewhat open-ended model if they do not have a positive history with this model or have not yet established a trusting relationship with you or your studio. If you wish to use this model, you should be prepared to at least provide a ballpark estimate of the project for the client as well as explain why this model benefits both the client and studio. If you are not able to comfortably and confidently have this discussion with a client, the client may insist on using a flat fee model and protect his or her own interest.

Hybrid Pricing. This model is basically a catchall for any other type of arrangement. Just as every project is different, so is the relationship between the videographer and client. What you are willing to do in exchange for a particular benefit for you is truly a personal decision. When we first launched our studio, we had a very strict payment policy and held every client to it. We needed cash flow to not only make payroll but also to expand our equipment and offerings. We did not barter; we did not trade anything except "checks" for services we bought or rendered. While it was something we needed to do, looking back many years later, we probably lost a few clients or discouraged clients from utilizing us for additional services because we were obsessively focused on our bottom line. Obviously, you need to protect yourself not only financially but also legally with every contract or interaction. Should a client propose an alternate payment exchange, take a few minutes to consider whether the alternative makes sense. It may have other non-tangible benefits that are less than obvious.

Lastly, the phrase we use the most during the proposal and contracting phase of any project is "We only like surprises on our birthdays and Christmas, and we promise not to surprise you on any day other than your birthday and Christmas." We don't believe there should ever be a shock or surprise when a client opens an invoice. We also don't believe we should ever get a phone call in which the client states they are not prepared

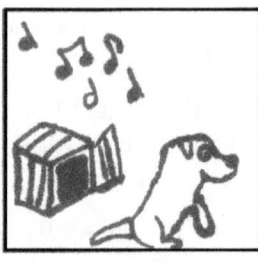

Standard "Royal Jerky Intro" with
Music Playing & Dog is Released.

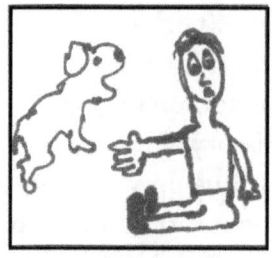

Dog jumps on Tommy and knocks
him to the ground.

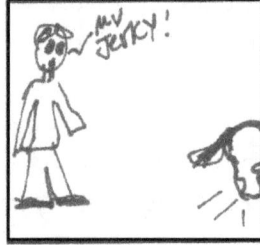

Dog takes Jerky from Tommy and
runs off camera

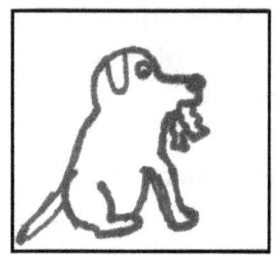

Cut to Closing Sequence with
Dog eating Jerky.

Figure 6.2
Every project will have its own unique story line.

to pay our invoice. Having a solid contract, a clear project plan, and regular communication throughout the project will greatly reduce surprises for all those involved.

Planning and Scripting

If you are a creative person, the two specific areas that you will migrate towards in the video production process are the planning and scripting phase and postproduction editing. In the planning and scripting phase, you should be working with your client to define the story. Depending on your type of production, sometimes this is predefined for you. For example, if you were retained to capture a sporting event or community function, the story will unfold before your eyes. There is nothing to be developed or defined. If you are working on a promotional, instructional, or entertainment production, extensive work is usually required for an effective shoot. For the remainder of this chapter, we are going to take the position that we are developing a production that is not capturing a sporting event or community function.

Each production or project will have a different storyline. You may be promoting an item, persuading a viewer, or simply providing some form of entertainment. Understanding the objective of the story is the first step. Once you have established the goal of the production, it's time to turn on the creative juices and create the messaging. Depending on your goal, one of your productions may require a very technical or specific message, while others may be less formal and more entertaining.

Once the storyline is in place, the next step is to identify the visuals required. Your video, graphics, voice-over, and any overlays should be consistent with the storyline and further enhance the user experience. Depending on your workflow or the expectations of the client, some serious time may be required to complete extensive storyboarding with text, drawings, and/or diagrams. Do not get caught up in the belief that complex story boards are the way to go; in reality, how you develop the story and get it approved by your client is greatly dependent not only on the experience you and your client bring to the table, but also the relationship you have. Some clients or projects may only require a basic outline.

If you are working on a new client's first project, you are much more likely to spend additional time documenting the story and formalizing a plan than you are with a client from whom you have completed many projects. Over time, as everyone gets to know each other, some of the formality may be eliminated. We want to be very clear; this does not mean you can skip planning a production or clearly defining your story up front. However, you may be able to reallocate the hours of labor required to produce complex storyboards or a lengthy written plan to other parts of the production.

Once you have developed your story and decided on the visual elements required for your production, it is time to select a location. Depending on your project, the best location for your shoot may be in a private setting, such as the client's business location, or a specific or specialized location may be needed to convey the desired messaging of your story.

Unless you are intimately familiar with the selected location, we recommend you conduct a site visit before the actual day of the shoot. Once you are on location, look around. Where do you anticipate the shoot will occur? Is the location ready for filming? Is the location clean? Or is the location cluttered with objects that need to be removed to facilitate a successful shoot? For 360-degree video, remember you are shooting in a sphere. The days of moving something three feet to the left or right or repositioning the camera to change the field of view are gone. Make sure you evaluate your location with this in mind.

While you are on location, make sure you have a conversation with your client or the property owner. Are you going to need a location release? Do you need a specific permit for the shoot to occur? Are there any privacy or security concerns? If you have all the Federal Aviation Administration requirements in place and are planning to use a drone, do you have clean air space, or are you too close to a restricted zone? If you are in a healthcare setting, is any specific patient or identifying information within your field of view? Gaining as much information as possible about your location before the shoot will improve your efficiency on the day of the shoot.

Once you have the story and location, you need to identify who is going to participate in your video. While certain productions, such as virtual tours of a home or technical how-to videos, may have very few (if any) participants, most videos are much more interesting when a variety of participants are involved. It's important to make sure you have the right participant for your specific production. At times, you may require a professional actor or actress, and for other productions a "real person" may be the right fit for the specific shoot. Whether you select professional talent or use a "real person," you should consider a few basic attributes.

Make sure the individuals act appropriately on camera. For example, if you are working in a healthcare setting, make sure they are following proper protocols and policies with privacy, personal protective equipment, and similar items. If you are working with a construction crew, be sure they are wearing their helmets, safety vests, appropriate footwear, and other required safety gear. If you are working in an industrial facility on a production line, do the participants have appropriate eyewear, clothing, and footwear? This is important. If your client ever has an incident and becomes the subject of an investigation that leads regulators or a governing body to review the video, it should help the client. It should show all the participants doing the right things,

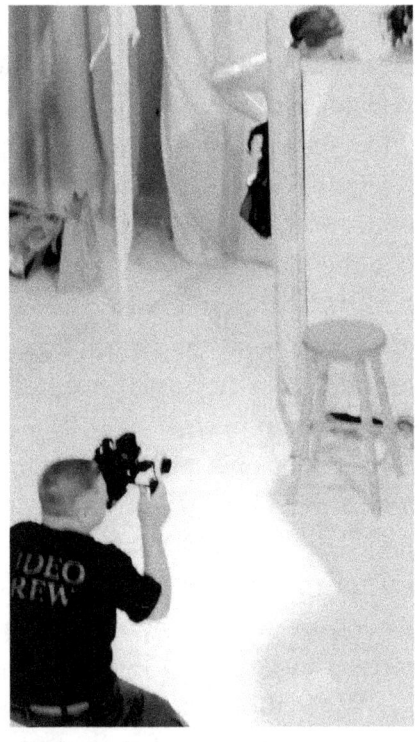

Figure 6.3
Participants should be dressed appropriately on all shoots.

wearing the right protective equipment, and following all the rules. It should not help establish a pattern of doing something wrong. Discussing this with all participants before the shoot is essential.

Without turning this into a legal, moral, or ethical conversation, there are simply some facts that cannot be ignored. Not every person will make a good participant in every production. You should consider each participant selected for your production. Let's look at a few examples.

You are retained to create a commercial for a high-end luxury car company. Upon your arrival, you are greeted by the client, who introduces you to his or her nephew and informs you that he will be taking the role of the main participant featured in this commercial. He is twenty years old and is wearing a T-shirt and worn jeans, with his hair dyed purple. Is he the best candidate for this production? Does he represent the image you wish to convey of the brand? Is he part of the target audience, or is he someone the audience can relate to?

Consider a scenario in which you are retained to create an educational video for a hospital. The proposed participant, who has the part of a leading expert in the script, arrives unshaven, hands covered in grease and eyes bloodshot as if he had been up all night. Is he the right choice to represent the hospital and its brand? Does he instill confidence or convincingly portray the role of an expert?

Reflect on a project in which you are going to demonstrate a highly technical process in an educational video. The participant selected does not have the ability to pronounce the technical terms, nor the confidence to be on camera. Will he be able to deliver clear, concise, appropriate direction with an authoritative tone to provide direction?

What about meeting a young lady who is designated to be the spokesperson for a family restaurant commercial but is dressed provocatively? Does her appearance align with the image of the family restaurant, or does it align better with a bar or nightclub?

It's very important to understand that you should never refuse to work with any subject based on their appearance, education, sex, age, religious beliefs, sexual orientation, ethnic background, race, or similar characteristics. At the same time, you do have a responsibility not only to yourself and your client but also to the individual. Although on productions you will specifically use talent for shock value, a good guideline is to never use participants in a production if they will not only become a distraction or an unsuccessful participant but also be put in a disadvantaged position. It's important you take time early on to evaluate and coach the participants before the shoot.

Your plan should also identify a definitive approach to the acquisition of audio for your production.

Will your participants be scripted, or will they simply talk on camera? While we believe that on occasion teleprompters may be used as an aid to convey very specific messaging or technical information or to support a lengthy production, dynamic or simply spoken lines without the use of a teleprompter usually provides a much more realistic and smoothly flowing recording.

Are you incorporating a voice-over into your production? Over the years we have used many different sources for voice-overs. As with anything, we've had some great experiences and we've had some disastrous results. You need to identify whether a voice actor or actress has been predetermined.

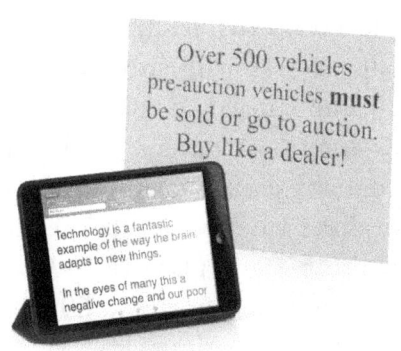

Figure 6.4

There are several inexpensive teleprompter apps that may be used to assist participants with the delivery of complicated messages.

If you can make or shape the decision regarding voice-overs, it is our recommendation that whenever possible, professional voice-over talent be used. It will improve the quality of your production, and will also cut down on the production time and editing of your final production. If you are forced to use local talent for your voice-over, local radio and TV personalities may be a viable option for you. Finally, if you're working on behalf of the owners and they wish to do their own voice-over, we recommend you spend a little bit of time coaching them. If they do not have a passion for their business or it's not easily recognizable through their voice, it is your responsibility to do everything in your power to lead them to a professional.

Most video productions will incorporate a soundtrack in an opening or closing sequence, or perhaps background music during a specific shot. Some clients will have an audio track that they use regularly as part of their branding strategy. Others may have prerecorded jingles. Some projects may require you to assist the client with obtaining new soundtracks. There are many royalty-free options available for free or at a very reasonable cost. Any planning you can accomplish in this area will help your editing crew on the back end.

The last part of planning is to identify and understand your distribution strategy. With traditional video production, there are many different options for delivery, including computer files on disk or delivered via the cloud, CDs or DVDs, or uploading to proprietary platforms. As discussed previously, at the time this book was written, 360-degree video had significantly fewer distribution options. It's essential to have this conversation early on with your client, not only to establish delivery expectations but also to obtain this information for the crews responsible for the shooting and editing process.

Setup

We have covered many of the specifics of setting up cameras and positioning gear in the previous chapters. Here are a few recommendations regarding the setup of an effective video shoot.

Be prepared. Each shoot is different, and you'll be working with different people. You will only have one chance to make a positive first impression. As a videographer, you need to be prepared and ready to work. You need to be well rested, well-groomed, and dressed appropriately. Our studio has gone as far as ensuring our crews are all dressed alike so we not only convey a professional demeanor but also are easily recognizable at a production shoot. We also have variations in our company attire that allow us to be dressed appropriately when working in a variety of environmental conditions and temperatures.

Your equipment must be ready to go. The day of a shoot is not the time to find out that you are missing a locking screw on the fluid head tripod, that the wireless microphone wires are frayed, or that the battery in your camera needs to be charged. Your memory cards should be clean and formatted. Your camera settings should be adjusted in anticipation of your specific location and lighting. Your gear should be packed and organized so it's ready for quick deployment as the shoot progresses.

Be early to the shoot. If you are not early, you are late. Have your gear unloaded and be staged to execute your plan before the shoot begins.

Bring extra gear. Be ready to switch out broken equipment or incorporate something different if there is a change in plans or if an unexpected opportunity presents itself. Having extra gear on the shoot also instills confidence in your

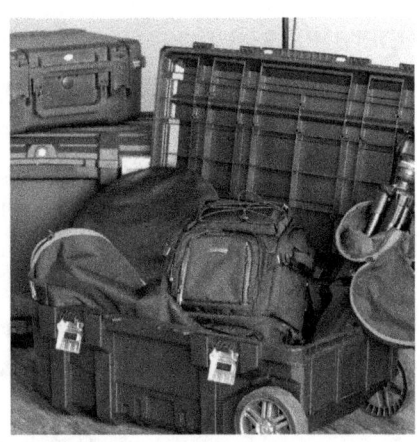

Figure 6.5
Having extra gear on the shoot also instills confidence in your client.

client. Put yourself in the position of a bride hiring a videographer for her special day. Do you think she would be more relaxed with a single videographer who brings a camera and a tripod for the shoot, or the crew that shows up with multiple cameras and backup equipment? In our early days, we used to carry an extra shipping container with 50 pounds of unusable, outdated equipment. Although it looked like we had extra gear and were prepared for anything that was thrown our way, in reality we were just starting out and didn't have the money for backup gear. That outdated gear in a heavy box has since been replaced with a cache of state-of-the-art equipment that we occasionally use on a shoot, and it's always visible to our clients.

The last step in the setup phase is the setup of the crew's state of mind. Do not allow anyone on the crew to become complacent. Every shoot is different. Stay alert, stay focused, and keep your eyes on everything around you. From an innocent bystander walking through a shoot to a change in weather or an equipment failure, you are not going to be able to accurately predict what will happen next. If you are alert and prepared, you will be much more successful over the long term.

Operations

Congratulations. You successfully completed a proposal, have wrapped up your planning, and are set up and ready to go. It's now time to complete your actual shoot. You have your plan and have your gear on the ground (or in the air), so you're ready to push the record button to start acquiring your video.

The inexperienced videographer will simply press the record button and call for action. If you follow your plan, you'll get good video, and on occasion, you'll get exceptional video.

Experienced videographers will understand that there is more to operating an effective video shoot. It's important to point out that many videographers believe that since we shoot on digital media, the cost of shooting video is basically free. This could not be further from the truth. While we no longer pay for film and processing, there is a cost associated with the time to acquire, sort, and process the footage. For this reason, it is just as important for videographers today (like their predecessors) to take the extra time to ensure they have an efficient and effective operation set up before they begin to acquire footage.

Have you checked your camera? Is it properly positioned to get the desired footage or field of view? Verifying the appropriate framing involves reviewing the entire sphere with 360-degree video; with traditional video, remember the traditional rule of thirds. Have you looked all throughout the entire field of view of your device? Have you looked through the entire sphere of your 360-degree video camera using the live preview mode? Now is the time to discover the soda can that was mistakenly put down in the middle of your shoot instead of spending hours trying to remove the can during postproduction. Is your lighting correct? Do you have enough power? Do you have enough data storage space?

Are you using multiple cameras? In addition to checking them all as above, are they positioned appropriately? If you are using multiple 360° video cameras, are they properly positioned in the blind spots of each device so neither camera shows up in the sphere of the other camera?

Have you checked your audio? Are the levels on all of your microphones adjusted appropriately? Do you have enough power for the microphones, recording devices, and mixing board if you're using one? Many microphones, mixing boards, and recording devices have several sets of input and output jacks; are your cables connected correctly, and have you verified that the recording will include all desired input devices and channels?

Once you're confident in your final checks of your equipment and setup, the next step is to answer a simple question: do you rehearse or not rehearse? Depending on your specific project, the participants involved, or the movement within a shot, we suggest you consider a brief rehearsal. With that said, you never know how a rehearsal is going to go. Participants may pull off an exceptional performance on the first try. For this reason, we recommend that you record each and every rehearsal or movement through the script. All participants should be encouraged to run through all lines and movements or other actions with the same focus and desire for a successful shoot as if they were participating in a final take. While we have recorded and used the first rehearsal run-throughs in a finished product, this is not typical. It usually takes four or five takes to generate the final product. Obviously, using professional talent almost always results in fewer takes and less production time on the set. Some participants may need additional attention and coaching for you to achieve the desired footage. Initial use of cue cards and clear marks for the working boundaries, sweet spots, or individual movement safe zones on the floor may be helpful to all participants, but especially to the non-professionals on the shoot.

Until now we have not spent a lot of time discussing movement during the shoot with 360-degree video. Capturing video in traditional video acquisition is relatively straightforward. Videographers have two distinct choices: they may place the camera on a static or dynamic mount and capture the movement within the field of view. With 360-degree video, it becomes a bit more complicated. With the limited field of view of a traditional camera, we can plan the shot and the movement of the camera to control not only what the viewer sees but how it is seen. With 360-degree video, we are moving a sphere. While we can define a default field of view or guide a viewer throughout the movement, the reality is that we have no definitive control over what they will look at or in what order they will view the content.

In a static sphere, the videographer can take reasonable precautions to ensure that only appropriate content is delivered. In a moving sphere, it becomes significantly more complicated (if not impossible) to guarantee the appropriateness of the content. Obviously, within a controlled atmosphere (such as a virtual tour of a home or business or working on private property) it is possible to produce a product with reasonable and often acceptable results. When you are working with a moving sphere in a public place, however, it's almost impossible to predict all that will be captured. We will cover the postproduction process in detail in chapter 7. For now, it is important to note that the editors of 360-degree video have the responsibility to ensure that the video captured in a sphere (static or moving) is appropriate for distribution.

In chapter 5 we talked about the importance of data management. An effective operation must incorporate data management into the plan and be ready to go before starting to obtain footage. On our smaller or single-camera projects, this responsibility often falls on the individual camera operators. They are usually responsible for not only acquiring the footage but also securing the original data storage card(s) and creating a backup of the data. On a larger shoot, we will designate a team member who has the sole responsibility of managing all camera, audio, and client-provided data. A detailed log and check sheet are used to account for all data and what processing has occurred. On-site backups are made, and that person is also responsible for the security and transportation of that data back to our studio.

Break Down

So, the shoot is over. Just like the end of any workday at any company, at the end of the shoot, most participants, clients, and bystanders will be quick to pick up their personal belongings and head for the exit. As a member of the production crew, you do not have that luxury. A few other items must be addressed before you call it a day. Again, some of you may find this information quite basic, but because I never cease to be amazed by what we see from other production companies or partners on our larger projects, I think it's worth repeating.

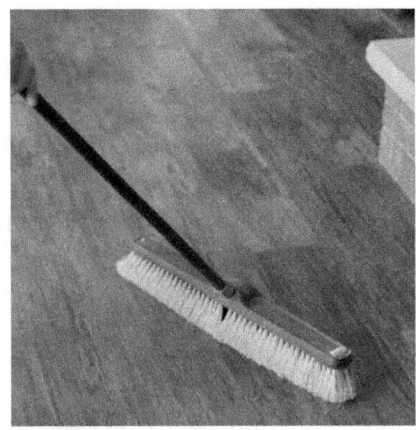

Figure 6.6
When clients see you sweeping the location on the way out, it shows that you care about them and the work you produce.

It's time to pack your gear. Take your time. By doing so, you will not only protect your investment, but you will also eliminate the need to handle your gear twice by having to repack it when you get back to the studio. Put all gear in its protective cases. Pay attention to the details, making sure things such as lens caps, retention straps, and traveling clasps are in place. Coil and wrap your audio, power, and data cables as if you were getting them ready for your next project. I am amazed by the number of people who simply wad everything up into a ball and stuff it into the bottom of a duffel bag. Make sure your data devices are protected and secure.

Make sure you clean up after yourself. You should always leave the location of the shoot better than you found it. Make sure all markings, any tape on the floor, or any other item you introduced to the environment leaves with you. Anything you can do, such as emptying the garbage can, picking up production litter, and returning items or fixtures to their original locations prior to the production, may not be noticed, but it is certainly better than getting a call the next morning about a noticeable problem from something left behind.

Help others pack their gear or assist them in the removal of their gear from the location of the shoot. We are very particular about how our gear is packed and will often decline assistance in the packing of our equipment. However, we are always happy to accept assistance with getting our gear to a loading dock or onto one of our vehicles. This is an area where a little goodwill can go a very long way, and long-term relationships and friendships are born and fueled.

Double-check everything before you leave. Make sure you have all of your gear. Make sure it is packed appropriately and secured in your vehicle. Make sure the property location is clean. If you're responsible for shutting off the lights, turning down the thermostat, or locking the doors or gates to the property, make sure that this is done. It is not unusual for us to be the last team to leave. If the owner or manager of the location is not with us when we are leaving, we also generate a quick email documenting our actions, notifying them that we are now leaving the property and letting them know the condition that it is being left in. We also thank them once again for the opportunity and their continued business.

For most production companies, this will be the end of the shoot. For us, there are two additional phases.

We complete a quick critique of the shoot. Depending on the project, this might be a formal review, but more often than not, it is an informal conversation with the crew and management team to review what went well and what could've been improved, as well as suggestions on how to improve our workflow moving forward. Sometimes these critiques reinforce some of our initial suspicions about the potential pitfalls identified during the planning process. They may also reflect how we dealt with an unanticipated event or action on the shoot. Learning to improve is a continual process. Even if you are a single videographer company, not critiquing your shoot will not only shortchange you and eliminate an opportunity to grow and develop, but also hurt your clients over a period of time.

The last phase or part of breakdown is a celebration. It's important to take time to celebrate your success. It may be as simple as saying "thank you" to your fellow crew members and gathering briefly over an adult beverage, or perhaps it may be a more formal occasion following the completion of a larger project or production. Although a celebration is certainly not required, our team not only works together but plays together, as well. We believe that the celebrations have made us not only a more cohesive team, but also a more effective and efficient one, which benefits our bottom line and ultimately the products and services we deliver to our clients.

Chapter 7

Postproduction Processing

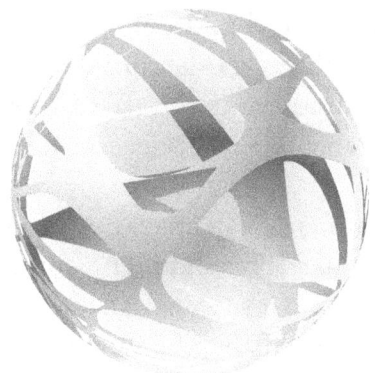

During postproduction processing, all the acquired assets (such as video, audio, still photos, graphics, and text overlays) are assembled to create a final product to achieve the goals of the project using the defined project plan. Some studios refer to this as editing, or simply "post." No matter what you call it, depending on the assets obtained, it can be the most rewarding or most stressful part of the project.

Before we go any further, let's talk about what this chapter is about and what it is not about. In the following pages, we will share an outline of our postproduction process. We will briefly touch on a few general principles and offer you some factors to consider when editing 360-degree video. We will not provide you with a step-by-step procedure or with instructions on how to use any specific editing software or utility. With the evolution of software and the cloud distribution systems being used today, such a tutorial would become outdated very quickly. We would rather provide you with a conceptual foundation that you can apply to any workflow and the software platform of your choice.

We value and put a great deal of emphasis on the planning process throughout all of our projects. We take the time to develop a postproduction plan for each project. While generally the plan follows our typical workflow, every project is different. At times, we will need to build or insert extensive animations or graphic overlays, while on other projects, we may not wish to incorporate background music for effect. Spending a few extra minutes up front can prevent unnecessary work or save hours down the line.

The first step for any effective postproduction plan is to define the final product. This should be completed before starting the editing process. In the next chapter, we will talk about the distribution process in depth. Identifying the distribution plan will clearly dictate or influence many of the decisions made during postproduction. Understanding the distribution format will identify the required export process and will most likely influence many design elements and the use of special effects, as well as define the appropriate safe zones used during the editing process.

For example, assume you are creating a product that will be distributed via Blu-ray disc. The resolution required of your video as well as the graphics and text overlay will be significantly different than if you were creating a low-resolution video to be distributed on a website intended for mobile devices. Once you've defined the required video format and identified the required resolution, file format, and possibly file size, you should define your data and back-up plan for the editing process.

There is absolutely nothing worse than spending an entire day editing video and then losing all your work because of an unanticipated event. Hard drives failures, unanticipated power surges, and operating system crashes never occur at the beginning of an edit. They always happen after several hours of work have been completed but have not yet been saved.

Before you even start to import your files, make sure you have a solid data and back-up plan in place. We typically keep a master set of files on our server as archive versions that are not used during the editing process. Copies of these files are then placed on the hard drive of the primary computer that is being used for the editing process. Next, we always confirm that no matter what software package we are using, its auto-save feature is enabled. Depending on your software, you may also be able to auto-save multiple copies or versions along the way. If you do, this is a very helpful feature to have enabled. If you don't, we strongly recommend that you do this on your own. When someone walks past our editing suite, it's not uncommon to hear a cell phone timer going off. While our newer team members may think this alarm signals time for a break or to meet the delivery man for lunch, it's really reminding the editor to take fifteen seconds and manually save the project as another version.

At least twice a day, the working files on the primary editing computer are also backed up to our server. While this process may seem like a lot of unnecessary effort, unfortunately we've had to use this file structure more than once following the catastrophic failure of a computer in one of our editing suites.

Once you have your data plan in place, it is time to start the editing process. Before we take you through the editing process, we would like to briefly address three general guiding principles that contribute to our success on a daily basis.

To Cut or Not to Cut?

New and experienced video editors alike will commonly face the question, "should I cut this shot or include it in the project?" A good guideline is if you're asking this question, it's probably something that does not need to be included in the production. Today, viewers both young and old are faced with many things that compete for their time and attention. In addition to their activities of daily living, they are distracted by phone calls, email, and social media. Most people simply have a lot going on. When it comes to video, unless the shot or segment is clearly providing a benefit to the production, is required to convey a specific message, or is clearly advancing the project to the desired goal, it probably should not be included. Again, if you are asking the question, it's probably something that should be cut.

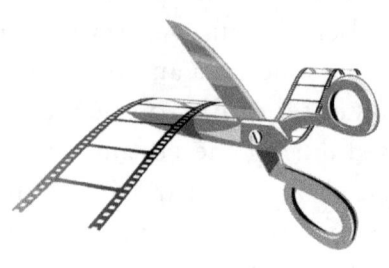

Figure 7.1
If you need to ask if something should be cut, it most likely should be.

More is Less!

For many years, we have been exposed to the phrase "less is more." While this phrase has a very similar meaning, we would challenge you to look at it a bit differently. As we just discussed, there's a lot of competition for a viewer's time. Simply including an extra shot that may show off the studio's capabilities or something not relevant to the production's goals may not be in your client's best interests. Like with anything in life, at some point, we are all saturated. We've heard enough, we've seen enough, and we either are ready to make a decision or have just plain lost interest in the topic. As technology gets faster and more lives are impacted by technology, the attention span of viewers continues to decrease. Video editors should strive to tell the story in the most efficient, effective, and engaging way. Unless you are producing a spot with a designated time interval for broadcast TV or some other medium requiring a specific time interval, your project should be edited to the shortest possible length.

Using Toothpaste as Spackle

If you've ever gone to college and lived in the dorms, this analogy will be easy to understand. If this situation does apply to you, hang in there; this will all make sense shortly. In college, I had a dorm room in an older building and had concrete block walls for three of the four sides of the room. The one remaining wall was sheetrock and plaster, and painted white. Back in the day, reusable adhesive mounting strips for hanging posters and photos to walls had not yet been developed. Anything we wanted to hang to decorate the room required a thumbtack or nail in the wall. While we were aware of the policies prohibiting the use of hardware that would leave holes in the walls, we wanted our rooms to look cool and feel like home. When we went home at the end of the school year, everything was taken off the walls. One of the last things that we did before leaving was to patch all the holes from the unauthorized wall hangings. If we did this and passed inspection by the resident advisor, we would get our security deposits back. If we didn't pass the inspection, within a few weeks we would receive a several-hundred-dollar invoice for patching and painting the walls. Thankfully, there was an easy fix. We didn't need a trip to the hardware store for spackle; we just used toothpaste. The toothpaste would fill the hole and dry crusty white, providing the appearance of a clean wall. Toothpaste was easy to work with, easy to clean up, and very inexpensive.

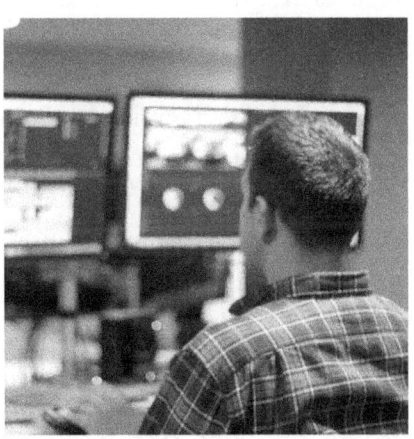

Figure 7.2
With a little patience, an experienced editor will find an alternative solution to even the most challenging problems.

Any experienced video editor will share stories of feeling boxed in the corner or stuck during a specific project. Like the concrete block walls of my dorm room, there will be several aspects of the project that cannot be changed—the angle of the camera, the action of the participant, or the distraction that may occur within the background of a particular shoot. It may take a second set of eyes to bring a fresh perspective to the project, but there is always something that can be adjusted to make it better. It might be as easy as using a different piece of footage, changing a cut point, or adding a transition, graphic overlay, or other manipulation. Often you can address the largest issues with a relatively simple, easy, and inexpensive fix. If all else fails, take a break and go get a tube of toothpaste. Along the way, you might come across the solution or inspiration to keep you moving forward.

In appendix C, we have included an example of a typical postproduction workflow. Following the diagram, you'll find a brief description of the content within each box identified as part of the editing process. Over the years, some of our interns have used a flow chart as a working project guideline. We have specifically broken that out of this chapter to allow you to do the same.

Editing starts with the collection of all assets. Although you will obviously include everything you captured using your cameras and audio equipment, your assets will also include any graphics, text overlays, soundtracks, voice-overs, or anything else provided or specified by the client.

Once your data is collected, the next step is to set up or organize the data according to your data management plan. It might require a few extra minutes of your time, and you may be tempted to skip this step, especially on smaller projects. We advise you not to skip the step; always organize your data. Should anything ever go bad with one of your projects, the time spent on developing this habit and all the time spent on this step for all of your other projects will make this all worthwhile.

Next, select the editing platform you will use. Like any company, we have a preferred platform that handles most of our workflow, but in certain cases, that platform just won't do. In the case of 360-degree video files, some proprietary software packages bundled with the cameras have unique features. These specialty features are still not available on some of the traditional editing platforms. Understanding the final project requirements or specifications may also influence the editing platform you select.

No matter which software platform you choose, the next logical step is setting up your project timeline. We typically import all assets to the timeline in one swoop. Then we organize them and configure the specific visual and audio layers to create a timeline.

The next thing we do is synchronize the audio files with the video content. Some platforms have this capability within the software, while others require a third-party plug-in or application to complete this task.

Once your video and audio files are synchronized, it is time to lay in the voice-over. Here is where the project starts to take shape. The entire timeline should be adjusted to reflect the voice-over, audio, and video placed on the timeline. To do this, the editor should start making preliminary cuts. The editor may also be adjusting the order of the sequences. In our world, it is not uncommon to acquire video out of sequential order. For example, depending on the production schedule, we may obtain the closing sequence before the opening sequence. This is when the detailed log kept during the operational phase of an effective shoot becomes an asset. Unless the editors were involved in the acquisition, they may be required to spend additional time locating the footage required. This process continues until the key sequences are trimmed and aligned with the production timeline. At the conclusion of this step, the rough draft of the production should be complete. At this point, you should render the timeline for the first time. Depending on the length of the timeline, the amount of footage, and the number of cuts, this rendering may be relatively quick, taking several minutes, or may take a few hours to complete.

After the initial rendering, the soundtrack or background music is added. Now is the time to make any adjustments to the timing of the audio track with the existing assets on the timeline for purposes of emphasis or effect. After these adjustments, the timeline should be rendered again.

At this point, you should add B-roll footage as required. The B-roll is any supporting or interesting footage that will add to or enhance the storyline while filling in any areas where a visual opening exists. It is important that you remember "more is less" and use a critical eye when selecting B-roll. The shot might be cool, but that doesn't mean it will be applicable to the final production. After the addition of any B-roll footage, you should again render the timeline.

Next you add any video or audio transitions between the cuts. Similar to B-roll, this is not an area where an editor should insert transitions to highlight their abilities or the studio's capabilities. The goal is to make the transitions from the cut segments almost unnoticeable. The viewer's focus should remain on the individual segments and not be drawn towards the individual transitions.

Any required graphics or text overlays are inserted next, followed by applying appropriate transitions to those as well. At this point, you should render the timeline again.

Finally, you should add any special sound effects or visual effects to create the desired response, and render the timeline once again.

At this point, the proof is ready for the client to view. Remember, you only have one chance to make a first impression. Even though it may be a first draft, it should be as close to the final cut as possible. On occasion, it may be appropriate

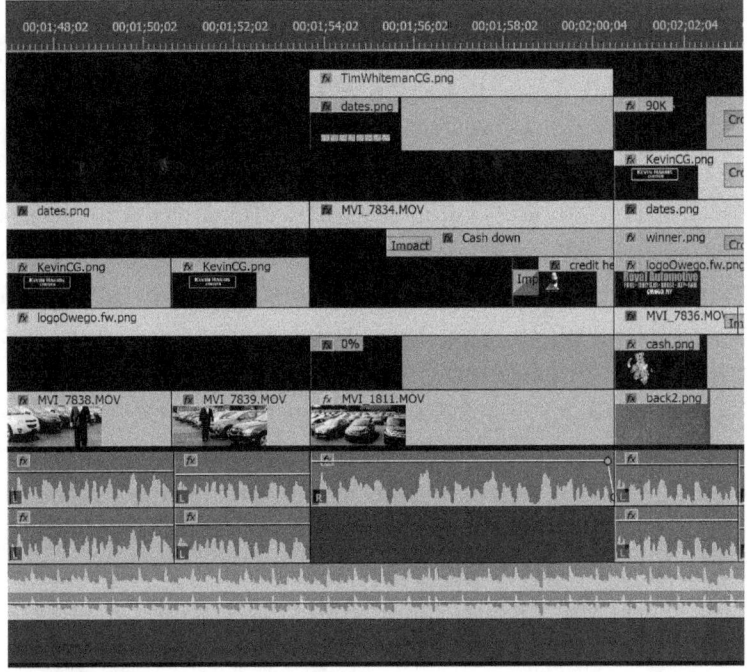

Figure 7.3
Regular rendering of a timeline will improve your playback during the editing process.

to provide a simple rough-cut version to a client to help with decision-making or obtain direction, but for the most part, all proofs should be as complete and polished as possible.

Should the client ask for any edits following review of the proofs, simply complete the edits as requested, render the timeline, and provide the client with an updated proof. This cycle should continue until the client provides final approval of the production and it's ready to be distributed.

Distribution should be relatively simple, because a plan for this should have been identified early on in the project. Depending on the requirements, the editing platform may be used to create a special encoded version for placement on a disc, the file may be exported for use on a website, or it may be uploaded to the specific distribution platform.

After the distribution plan is executed, the editor's work is not done. It's important that all production files be copied to a secure location from the main computer in the editing suite. Our studio also takes the time and commits the disc space to make a second copy of these working files to preserve them for potential future updates or edits for the client. As your company grows, it's important to identify that these extra steps not only add to the time for each production, but also have an associated expense related to the long-term storage space commitment on your server or within your data bank.

Now that we have provided you with an overall overview of the editing process, let's discuss some specific things related to the editing of 360-degree video. Again, at the time of the writing of this book, 360-degree video technology was emerging and changing daily. What we're doing today is different from what we were doing just a few weeks ago. Our software platforms are configured to continually receive new plug-ins and updates to handle this emerging technology. With that in mind, we can offer only a few specific pieces of helpful information related to 360-degree video.

As we said early on in this book, 360-degree video is obtained from many lenses and stitched within the camera. Most cameras produce a video clip that looks and acts very similar to a traditional clip within the editing suite. The editor will notice that the exported clips appear as a flat panoramic video of the entire field of view. This file is then delivered to the viewer in a sphere through the specific video player or distribution platform.

Similar to editing traditional video, the 360-degree video editor can manipulate individual clips or complete batch editing on a complete timeline of clips. Adjusting the size, brightness or contrast, hue or saturation, and curves, as well as adding noise, sharpening, live masks, or similar transformations are all possible with 360-degree video.

When editing 360-degree video, the editor needs to be familiar with one major difference. When you acquire a traditional video, your camera captures the field of view in the direction it is pointed. When you acquire 360-degree video, the camera's forward perspective at the center of the lens is the focal point within the sphere. Many of the proprietary software programs and a few of the mainstream editing platforms allow for this field of view to be changed. By adjusting this field of view, you can refocus the viewers' attention or guide them through the sphere. In traditional video editing, this process is similar to panning and then key-framing the video to move your focal point.

At first, this was a difficult concept to understand. I think this example will resonate with all editors. My wife and I went shopping while wearing our matching 360-degree video cameras on dynamic helmet mounts. Our mission was to find me a new sport coat for an upcoming presentation. As we walked through the store among the racks of clothing, our helmets captured our adventure.

My camera captured me entering the store and then going to the rack of sport coats, trying one on, heading to the register, and walking out. The footage was similar to a virtual tour. It flowed and was relatively easy to follow.

My wife's camera captured her entering the store, stopping at a rack of shirts a few feet from the door. She then walked a few feet, turned, and

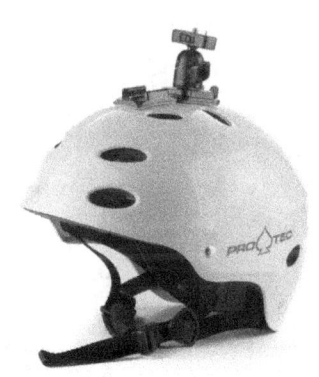

Figure 7.4
Repositioning the perspective within the sphere as the sphere moves is a skill all editors should master.

looked at a rack of pants. She walked a little further, turned the opposite way, and looked at a rack of neckties. Eventually she caught up with me as I was trying the sport coat on and accompanied me to the register.

Now consider the footage from her camera. It was choppy. She turned her head as she focused on the numerous racks. Her head was looking up and down, which raised and lowered the focal point within the sphere.

One of the exciting opportunities of working within a moving sphere is that you can reposition the perspective within the sphere as it moves. In short, we can take the second video clip from my wife's helmet camera and make it very similar to my clip through the editing process. Her footage may actually be more interesting to viewers, because she captured more content while getting to the same end point. For you nontechnical readers, I was told that the real difference in the footage acquired for this example was that I went to buy a sport coat, and she went shopping.

Advanced editing may also be done using 360-degree video, but the sphere creates some additional challenges. As we mentioned earlier, most editing software platforms will display the video in a flat panoramic format. If you are used to adding overlays, logos, or other graphics to cover something in traditional video editing, you can do the same here. This technique is useful for covering up the object that should have been removed prior to the start of the shoot, or perhaps a logo on a piece of equipment. We've also used this process to repaint a room or a logo on the side of a box truck, as well as to animate a logo or graphic within a specific video. Using an overlay or mask, the editor would simply utilize key framing to make sure the position of the overlay or mask either holds a specific position or tracks with the video. In chapter 5 we presented information and an image about using 360-degree video with a tripod. This is the technique you would utilize to remove or alter the presence of tripod legs.

While we have not yet done this, we have seen videos where editors have significantly altered the sphere by blacking out and branding a certain area of the sphere. We've seen this technique used to restrict the user from viewing something like looking through a window, looking at the ground or looking at the floor and ceiling. During our research for this book, we also viewed several videos that appeared to have been shot on camera platforms that provided a limited sphere of 270 degrees. The remaining part of the sphere was built in postproduction. The possibilities are endless.

On the surface, this sounds fun and relatively straightforward. Unfortunately, it's not. If you are working towards the middle of the flat panoramic, it's straightforward. As you get closer to the edges of the video, it gets significantly more challenging. Remember that the cameras are shooting using a wide-angle lens and that these are prone to aberration, especially near the edges of the field of view. Stitch lines created when turning the flat panoramic into a sphere will create a significant challenge as well. When you have these figured out, consider the bend that occurs when the stitched panoramic video is converted into the sphere for viewing. Even the most experienced editors will need to develop a new way of thinking and practice with this technology before committing to a big project.

If you're an experienced editor, you should catch onto editing 360-degree video relatively quickly. If you are a brand-new editor, you have an advantage because you will not be distracted or slowed down by figuring out what techniques from your experience of editing traditional video you can or cannot use when editing 360-degree video.

Part of our production team recently visited one of the large theme parks in Florida and had the opportunity to experience a ride that combined movement and panoramic video technology. As we left, it was surreal to listen to our team members describe how to apply the 360-degree video technology at a cost of a few thousand dollars in equipment and some editing time to replicate something that cost the theme park several million dollars to create.

Make no mistake, editing 360-degree video will take some additional practice and has a somewhat steeper learning curve, but with a little practice, the possibilities and rewards are nothing short of awesome.

Chapter 8
Distribution

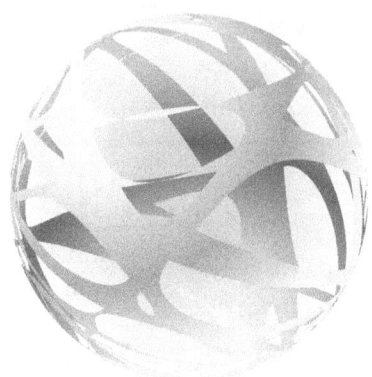

Distribution is probably one of the most important parts of any video project. The highest quality professional production will provide absolutely no benefit to the client if the final product cannot be viewed by the intended audience. When compiling all the information for this book, we left this chapter for last. Unfortunately, much of the content and specific details related to the distribution of 360-degree video are still being written by the industry and evolving every day.

As we did in the last chapter, we are not going to focus on specific distribution steps or the few specific platforms currently being used to distribute 360-degree video. We believe it is more important that you should understand and be comfortable with the theory behind a successful distribution plan rather than spend time learning a step-by-step process that will soon be outdated.

The most important part of the distribution plan is understanding your audience. To understand your audience, you need to understand the content they are interested in, how they are obtaining that content, and what devices they are using to access the content.

Look around you. Pay attention to the use of technology by your coworkers, friends, and family. Pay attention to those around you the next time you go to the mall. What technology are people using in the waiting room at your auto repair shop, doctor, or dentist? Look at the specific technology your parents or grandparents are using, and then look at what is being used by your kids and their friends. Take a minute and jot down your findings. Identify what devices are being used and what applications are being used. Do you see any trends?

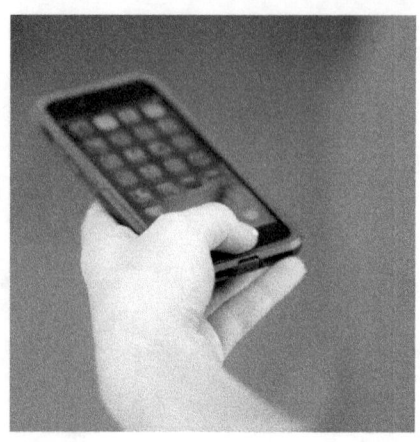

Figure 8.1

Understanding your audience's use of technology is an important part of your distribution plan.

Now look at the advertising campaigns directed at these groups of people. What are they seeing on their technology devices, and what are they being targeted with on the Web, on television, or in print? Who are the specific target audiences of these campaigns, what are the goals of the campaigns, and what technologies are they leveraging?

Every market is slightly different. Not every individual follows every trend. With that said, here are some trends that we have identified that may also be applicable to other markets today:

We have noted that the younger viewer is more active and involved in self-directing the flow of information. Much of their technology is delivered in a handheld format, with the smartphone often being their device of choice.

As audiences get older, their flow is less self-directed. Unlike the younger viewer, who is searching for specific information from a variety of platforms or feeds, the older audience member may turn on the television, open an email, or perhaps visit a favorite website. They may use the same technology as members of the younger audience, but they use it in a significantly different way.

Early adopters of newer technology are teenagers and middle-aged consumers. Although many factors influence early adopters, including access to the technology, household income, and influence by their peer group, these are the two primary groups that will be your target audience for integrating 360-degree video into your product offerings.

Once you have identified the trends in technology, you should take time to understand what content the viewer is interested in. For example, are they streaming music videos, watching movies online, or surfing the web? If they are active with social media, note which platforms they are using. It may be a challenge to identify the specific source of their content. It's important that you work with your clients to obtain their insight and direction. If they are successful, they should be able to identify and convey much of this information to you in a matter of minutes.

With this information in hand, it is time to align your distribution with as many of your findings as possible. If you're going after an older, more passive audience, it may be a viable option to have something that functions well on a desktop, is distributed on a DVD, or is part of a television campaign. If you're going after a younger audience instead of a middle-aged one, having something that can be distributed through social media and viewed on a mobile device is essential.

Over the last several years, the distribution of content has adapted to new technologies. By changing file sizes, compression rates, and other file attributes, content that once could only be delivered on a desktop computer or by a DVD is now served on the smallest of handheld devices.

Virtual reality and 360-degree video are currently forcing the evolution of delivery platforms. As we previously discussed, 360-degree video is basically presented in a panoramic format. Each individual platform generates a spherical video and provides the user interface. Some platforms use the device's sensors, allowing viewers to become more active in their delivery by simply tilting their phones or rotating their devices. This is the technology that will continue to be built upon for the delivery of 360-degree video and will ultimately enable a virtual reality response experience.

Although the specific integration of this technology is currently limited, we anticipate it will eventually allow for the full integration with desktops and perhaps be packaged for viewing on "smart" televisions. Other devices, such as virtual reality goggles, will also shape the distribution of 360-degree video. It's difficult to imagine all the possibilities for integration that exist.

How to Mitigate a Disaster When Something Goes Wrong

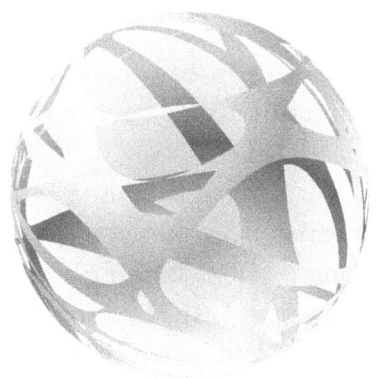

Whether you are a professional videographer retained to capture a once-in-a-lifetime event or an amateur videographer excited to be using video to share the story of your special family occasion, you need to be successful. Throughout this book, we have attempted to stress the importance of developing and executing a plan and provided you with strategies to help you succeed. In this chapter, we will talk about how to mitigate a potential disaster when something goes wrong. It is our hope that you are reading this chapter before you find yourself with a problem.

Hopefully you have already taken the time to develop your own plan for your project and have carefully implemented a strategy and, if you are faced with an issue, it will be something relatively simple or minor that should be easy to fix. We know from experience that even with the best plan in place, sometimes things just happen. We also know that sometimes life just gets in the way, a plan is not made, and for whatever reason, a crew simply "wings it." We also know that like our studio, many have videographers on call to capture breaking news or to capture footage for a client of an event or incident with very little notice. Sometimes there is simply no time to plan. We understand.

If you are using this chapter in the heat of the moment, before you try to address any specific issue, you should immediately take these three steps, no matter what the issue is:

Remain calm. Take a deep breath. Take a break for a few minutes and grab an appropriate beverage and snack, for you may have a long haul ahead of you.

Stop and think. What just happened, or what was supposed to happen that didn't? What are you trying to do and can't? What are you doing, and why it is not working? What are your options now? What are your limitations? How much time do you have? Is this something you can handle, or do you need help? Do you have enough money to get the help or purchase another solution?

Make a plan. If you recognized that you have a significant problem, it's best to develop a formal plan. If you simply react or start trying multiple things, you may make matters worse. The first thing we tell organizations who call us for assistance is to not change the power state of any piece of equipment or computer until you have a plan in place. Next, if you're having issues within your editing software, it may or may not be appropriate to save a copy of your work into another file now. Some systems are set up so that when you save the file, it removes the history of your actions that may have been saved within the program's cache or other autosaved backups that may be helpful to you. Do not ever simply hit the save or save as button.

Although we cannot address or cover absolutely everything that has happened to us or some of our clients who have come to us with issues, we have complied a list of some of the common issues videographers face. The information has been grouped by topic, but the topics are not listed in any specific order.

Talent

No matter how much preplanning you do, one area out of your control is the talent—especially if you had no input or involvement with the selection process. Over the years, we have dealt with participants who were intoxicated or uncooperative, showed up hours late, or didn't show up at all. There is no definitive solution for this type of issue. You will need to take one of two directions: continue the shoot as planned, or reschedule the shoot.

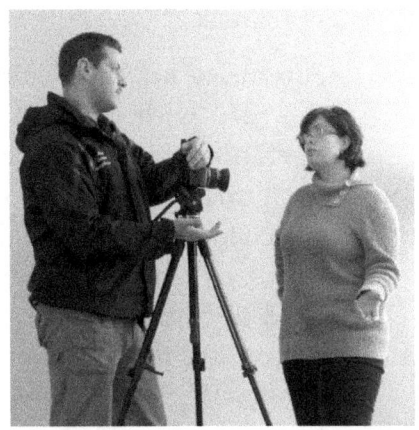

Figure 9.1
Problems with talent may require a new approach to your project.

We have been fortunate to mitigate potential delays and soaring costs by continuing almost all the shoots we have had scheduled. We have replaced professional talent with other participants who may have been affiliated with the client, family members of those on the shoot, and even one of our crew members. It can get tricky if you need to have them dressed a certain way, perform a specific skill on camera, or comply with a specific requirement (for example, the participant needs to be a 5'10" female with red hair who can play the tuba), but for the most part, it's usually much easier and more economical to keep moving forward. With some quality coaching and patience from all on the set, you will be surprised at what you can make happen.

If you are involved in the discussions about rescheduling a shoot or you have any input, you should make sure all parties understand the consequences of rescheduling. In addition to the obvious delay in the acquisition of video assets, rescheduling is likely to impact the editing schedule as well as create additional project expenses. Should the decision be made to reschedule, our suggestion is that this should not be done by the video crew for a few reasons.

- Any delay has the potential to create other issues down the road with the rest of the marketing plan, business operations, etc. You don't want your name affiliated with this.
- Rescheduling a shoot almost always results in an additional expense. In addition to your crew, others will need to commit to another shoot. You don't want to be responsible for this expense or put yourself in a position where the client is looking to you to be reimbursed for this expense.
- This is not only an additional expense for everyone else; it's an additional expense for you. You should be compensated for the time and any associated expenses that will be added to your production expenses for the shoot that is being rescheduled. It's a bit awkward and probably a little shady if you make the call to reschedule and then invoice the client for additional expenses.

This is one area where the client should not only be aware of the situation, but also an active participant in all related discussions and ultimately responsible for the decision to move forward or reschedule.

Camera Failures

You've set up your shoot. Everything is in place. The action begins, and a few minutes into the shoot, your camera stops working. Now what? First, we always recommend that you shoot important events, scenes, or occasions with a minimum of two cameras. Although some clients may balk at the expense or studio owners may argue against the necessity, it's a plain fact that having a backup copy of the video is always better than not having any footage should a camera fail. Cameras will fail. If your second camera is running, you have time to address the problem. If there is no second camera and this is a scheduled shoot that is staged for a promotional, educational, instructional, or other similar production, stop the shoot and regroup.

If you had only one camera running, grab your spare camera, get it set up, and get rolling. While it is possible to have a second camera fail, we've only had it happen once, for one of the reasons we are about to discuss. If you have the time, we recommend taking a few minutes to try to get your second camera working so you have backup footage for the editing process.

Cameras fail for only a few reasons. A camera that has been dropped during transit or during setup that has internal damage usually will not turn on or begin to record footage. It's very rare that a camera would start recording and then stop from being dropped earlier. What typically happens is something in one of three categories: power, storage, or temperature.

Most cameras today are powered by lithium ion batteries. Although they have a great charge rate, shelf life, and cycle life, when they get old, they can drop from 100 percent to 10 percent in just a few minutes. The first thing to check when your camera shuts off is the battery. Certain manufacturers also put a sensor within the battery door enclosure that will shut the camera off if the door is opened. Our crewmembers always carry several spare batteries for the cameras they are operating, and they are trained to immediately replace the battery in the event of a camera failure. Replacing the battery immediately eliminates any problem with a battery and cycles the battery door if that is the cause of the issue. On certain shoots, some of our camera gear may use an AC power source. We have been on shoots where a circuit breaker has been tripped or an extension cord accidentally unplugged, which resulted in a camera outage. Again, simply replacing the AC adapter with a battery will quickly alleviate most power issues.

It's important to have an empty memory card in the camera at the beginning of each shoot. Most cameras will automatically turn off or stop recording when a card is filled to capacity.

Like battery compartments, some manufactures also place sensors on the doors that protect memory cards. Should the door be open or not be seated properly, the camera will not function. Our teams also carry spare memory cards and insert a clean card into the camera if changing the battery does not alleviate the issue.

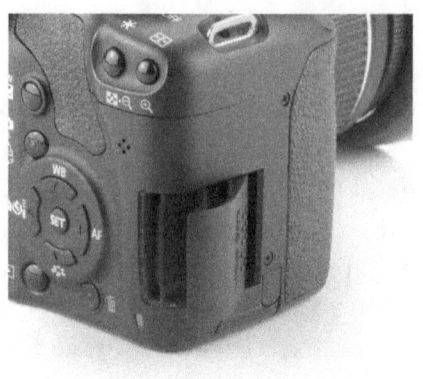

Figure 9.2

Something as simple as an open battery door may disable your camera.

While today's cameras have a wide range of operating temperatures, we have had cameras shut down because they were overheated. Most cameras have sensors that will trigger the shutdown to protect the advanced components within the camera body. It is not hard to overheat a camera. It will usually happen after several hours of use in a warm or humid environment. It may also be triggered if cameras are used in direct sunlight for an extended period of time. Over the past several months, we've been spending more time in the Southeast, and at least two cameras have shut down after extended filming on a pool deck in the middle of the day. Unfortunately, if an increase in temperature is the problem, there is no quick fix. You need to shut the camera off and remove it from any direct sunlight. Preferably, you should take it to a room that has an ambient temperature near 70 degrees Fahrenheit. If you take it to a cooler environment, you may fog the lens or other components, which we will discuss next.

Camera failures are the most significant issue a videographer will encounter. The other two issues that we see on a regular basis can be prevented. The first one is fogging or condensation on the lens. This fogging may also occur within DSLR camera bodies or waterproof enclosures. Fogging is caused by quick changes in temperature. We often experience this when going from a warm production vehicle to a cold environment, such as a hockey rink. We recommend that you wait for the fog to clear. It is better to let the fogging correct itself than to reach into the lens or camera body with any type of cleaning cloth or wipe. If you plan ahead and get to your shoot early enough, you can allow the cameras to acclimate to the temperature, and you should not see this problem. If it occurs on a shoot, simply wait it out.

We see one other issue on a regular basis—cameras coming loose from their mounts. There is absolutely no excuse for this. All cameras should be secured tightly on any mount used. If it's a dynamic mount, the camera should also be tethered. If you regularly work with certain pieces of equipment that have had previous issues, be extremely attentive to them. When a camera breaks loose, it usually causes significant damage. There may be damage to someone near the camera, to someone's property, or to the camera itself. Should this happen, your first priority is to make sure that no one was injured. Next you should worry about the damage that your camera caused, and last about the camera itself. Should serious injury or damage occur, our only advice is to obtain good legal counsel.

Lightning

We've also had one other issue that was power-related—supplemental lighting. Every time we have had an issue with lighting, it was one of three things: a loss of power because an extension cord became unplugged or a switch was accidently off, a dimmer box failed, or a bulb was simply blown. The fixes for these are relatively straightforward and do not require any further explanation. The loss of lighting should become obvious to anyone on site, and if taken care of immediately, the problem should not create any other issue within your project. Extra bulbs, extension cords, and dimmer boxes should be part of your spare equipment cache.

Data

In several areas of this book, we talked about the importance of having a solid plan for managing data at a shoot and during the editing process. Most videographers are protective of their data cards, and we are rarely called upon to consult about a data issue. When we are called upon, it's usually very bad. At least once a year, we will work with another videographer who has lost data.

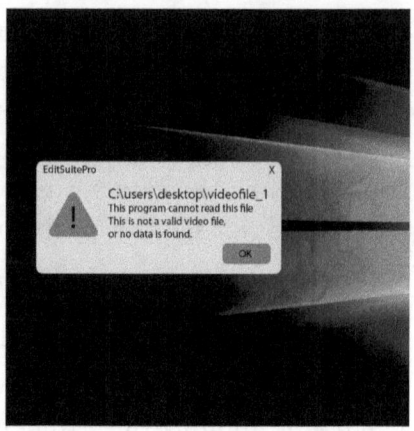

Figure 9.3

Data loss can be not only upsetting but also expensive to manage.

I know firsthand how upsetting data loss can be. Earlier this year, we migrated our studio data from one mass storage device to a newer, "more stable" platform. During this process, we moved over fifty terabytes of client data that spanned an eight-year period. Two weeks after the migration, we realized we were missing an entire client directory. In addition to several substantial video projects, we had also lost backups of their website, working graphic design files, and all our correspondence with them, which represented almost eight years of work. Fortunately, much of the data pertaining to their website, correspondence, and graphic design files was scattered throughout the cloud. We had DVD copies of their final products. We did lose several hours of B-roll and interview footage that was used on the DVDs, as well as other video assets. We were fortunate that much of what was lost was dated and probably not viable for future projects, but at the time, it made us sick. Many months later, it still stings!

If this happens to you and you call us for help, the first thing we're going to ask you to do is to give us a step-by-step account of everything that you did or did not do regarding the management of your data. We are not looking to find fault, get you to admit that you made a mistake, or make you feel worse than you are probably already feeling.

In our case, much of the data was years old, and our ability to recover any of the working files was just not an option. For security purposes, the device from which the data was moved was destroyed. Our only option, should we have needed to recover any of the video, was to contact the client and reshoot the video. This loss was much different for us than the loss of data that may occur in the middle of a current project, and you will most likely have several options for recovery.

Although our data loss did not affect an ongoing project, we did review the way we store our data. I guarantee if this ever happened to you, you would also review your internal controls. Do it now. Don't wait until something bad happens! If you are looking to safeguard your data further, the following paragraph should provide some assistance.

Split your data on multiple cards instead of using one card for an entire shoot. Change your cards frequently. Do not wait for the card to fill up. Should one card become corrupt or get lost, having your project split up may provide you with enough footage to still deliver your product. When you remove a card from the camera, immediately back it up to one or more sources. If you're transferring data using a camera, do not try to view any files during the transfer process. Always keep a full backup of your files in addition to those being used for an edit.

If you do experience a data loss, here are some basic things to consider:

Where are the cards that were used in the cameras for the shoot? Have you deleted files, or just made a copy of the footage from them? If you did delete the files, has any new footage been placed on the cards since?

You may be able to use a simple card recovery program. If so, that will cost you all of thirty dollars online and have your card restored within five minutes.

Did you back up the cards? To where? Did you make one copy, or several copies? Are they all missing? Are the files on your server? Is there a backup of the server you can pull the files from? Are they on the workstation? Is a backup of the workstation available? Did you make a copy for the client? Does anyone else have a copy?

Although losing your project files in the middle of an edit may cause a significant delay in your production, if you can recover the original project files, you can still complete the project.

If you implement only a small portion of our recommended data management plan, it is very difficult to lose every copy of your footage. Unfortunately, as our recent experience shows, there are those rare situations where the data is gone for good and recovery is not an option.

Traditional Video

If you chose to only use a single camera and realized when you attempted to download your card that your video did not record, depending on your project, there are a few ways to handle this. Can you repeat the shoot? If the production was promotional in nature, such as a TV commercial or a training video, you may be able to reschedule the shoot or hold the set over. If this requires you to incur an additional expense or pay for extra expenses related to the talent, you may need to reduce your fee. Although this may be a difficult way to learn the importance of using a second camera, it's a better alternative than not being paid at all.

If your event was a once-in-a-lifetime event, such as a wedding, a sporting event, or another family occasion, it's not something you can simply repeat. Your options become more limited. The easiest thing you can do simply is inform the client or your family that you are unable to produce the video and deal with the consequences. While this option may be the easiest, it is not what we would recommend. One viable option may be to obtain any cell phone video or still photos from other attendees and use a generic soundtrack. I have seen this type of production done intentionally. It has a very different feel and provides some challenges to the novice editor, but it is certainly a better alternative than simply saying "I didn't get it" and having to deal with those consequences.

The second most common issue we see with traditional video acquisition is footage from the camera that is either washed out or very dark. This problem can often be prevented by checking the camera's settings and playing back clips or using your camera's live preview mode before you start the shoot. Although there are many advantages to having your camera set up appropriately before the shoot, today's editing tools do a decent job of adjusting footage that may be washed out or dark.

360-Degree Video

In addition to the problems and issues that are discussed in this chapter, 360-degree video clips may provide you additional challenges in three areas: working within a sphere, editing spherical video clips, and exporting or uploading 360-degree video.

Early in this book, we discussed the importance of getting to know your camera and understanding the locations of your camera's blind spots and stitch lines. We cannot emphasize enough the importance of understanding these parameters before recording video.

What typically happens is that a videographer who is not yet experienced with 360-degree video will simply place the camera on a shoot and not spend the time to fully analyze the sphere before capturing video. When the video is being edited in postproduction after the shoot, the videographer finds that either something crosses partially into the blind spot and is out of place, or is creating an obstructed view, or there is significant aberration at a stitch line. An example of this would be a stitch line appearing in the middle of a banner, photograph, poster, or something with detail rather than being positioned on a blank wall or solid object.

If the shoot is complete or repeating the shot is not an option, it is up to the editor to develop a solution to this issue. Some editors will work their magic by using a blur tool or keyframing an overlay so that viewers will never realize what was done. If your editor does not yet have that skill set or the production budget does not allow for an extensive edit, rather than trying to blend something in to the area, it might be easier and have a better effect if the editor deliberate adds something like a graphic, text overlay, or studio imprint. Often deliberately inserting an edit such as this is more effective than trying to blend or repair at a stitch line. The same technique or principle may be an effective solution for addressing the issue of something crossing through the blind spot.

The second issue we see with 360-degree video happens when an inexperienced editor tries to edit the focal point within a sphere when the sphere is moving. In chapter 7, we used the example of shopping for a sport coat to illustrate the capabilities and effects that may be obtained when changing the focal point of a sphere with in a moving sphere. If done correctly, moving the focal point may be an effective way to guide viewers through a video while still allowing them the ability to independently control their view throughout the sphere. When done incorrectly, it will make the video choppy and hard to follow, or it may also create a feeling of motion sickness. There is no easy fix for this issue. Additional time and experience with the editing suite is required. If you are not an experienced editor, rather than take on the challenge of a moving sphere, you may wish to do a sequence of multiple static shots and develop a video sequence by using either a cut or fade transition between each shot.

The exporting or uploading of 360-degree video may take significantly longer in comparison to the traditional video that most editors are used to working with. We have had projects with only a two-and-one-half-minute running time take over one hour to upload.

We recently provided support to another team who did not realize the length of time this upload process would take. This person repeatedly interrupted the upload as it was being processed. The software created a set of temporary files for each incomplete attempt and stored them in the program's cache folder. The repeated attempts and cancellations were not done properly, which resulted in all the files being retained on the computer. This one project brought a state-of-the-art eight-thousand-dollar video editing workstation to a screeching halt. Do not wait until you have very little time left before your deadline to process your 360-degree video. If you need to stop an upload, use the software's recommended method of canceling the export or file upload.

With some platforms or software packages, an editor may not always be able to tell if the upload is progressing or if it has stalled. It is better to wait an extra hour than to have to spend several debugging a sluggish workstation.

Audio

Audio issues are usually straightforward and can be grouped into one of three areas. Following the shoot, you may find that you either have no audio, a low-level recording, or have captured something that you do not want in your recording.

If you have no audio recording, chances are you forgot to press record (yes, we've done that too), your recorder lost power, you forgot to put a memory card in the recorder, or the card you inserted was

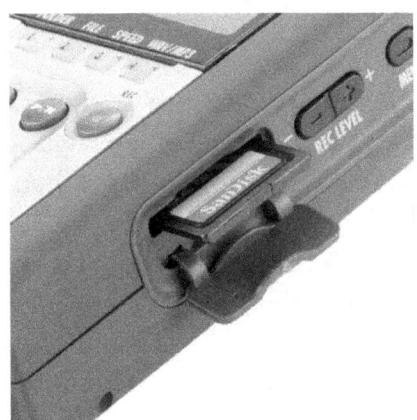

Figure 9.4

Checking your gear before your shoot for something as simple as a missing data card will prevent many issues.

full. If this happens to you, your first fallback might be using any onboard audio recorded by the camera. Although this is certainly not preferred, it may be your only viable option to use actual audio from the shoot.

If you do not have camera audio, you may be able to find similar ambient audio by using purchased audio clips from a royalty-free audio website to rebuild what you need. This is not a quick or easy process. To do this correctly, you will not only need to adjust the timeline of the audio to synch with the events captured within your video, but also adjust the volume of your audio track as well. Some other options that you may wish to consider include simply not using any audio at all for a specific sequence within your production or using a soundtrack with or without a voice-over.

Low-level recordings are usually the easiest to correct. Most editing software packages today can significantly increase the gain and/or apply additional filters to bring up your audio levels. If you do not obtain the desired results, an "old school" trick is to take an audio track and continually stack it on top of itself in a new layer on the timeline. This will also bring the levels up. We've had a couple projects in which we've had one track stack on itself four times before we had the desired result. It doesn't look pretty on a timeline, but it works.

What about when you captured something you really didn't want to record? When this happens, every situation is different. Sometimes your only option is to remove that section of the audio track. Here are two examples that may help you decide which action plan might be appropriate for your situation.

Let's say you are recording audio of kids playing in the park to advertise a summer fun program. The kids are running, jumping, playing on the swings, and just plain being kids. At a key point in your sequence, a truck drives by and backfires. If you are simply using this audio as ambient background noise and it's not timed to a specific event, chances are you that could simply make a cut and insert a similar section from another part of your audio track to replace the section with the backfire. These are easily blended with transition tools found in most software suites.

What do you do if you are recording an interview on camera and a nearby driver is heavy on the horn? First, from my perspective, this clip should never have gotten to post. If your audio crew was paying attention with a headset in place, they would've realized that the horn could be heard on top of the interview and should have immediately stopped the interview. A quick adjustment of the focal length of the interview and repeating the question would take seconds in comparison to the significant task of removing the sound of a horn from an interview.

If, for whatever reason, the problem wasn't noticed and the interview went on, unless it is absolutely critical not to do so, we would restructure the interview and simply do away with this part of the clip. If it is critical to include this section, we would be forced to use specialty third-party software to isolate the voice from the background noise and horn. Once the voices, horn, and noise were separated, the horn would be removed, and then the voice would be placed back on top of the ambient noise track and merged. Sounds easy? It's not. Depending on the number of variables, this process could take anywhere from half an hour to several hours of editing. Again, if this happens and you cannot repeat the shoot, find any way possible to simply avoid using this section of the clip.

Clients

Over the years, we have had the opportunity and pleasure to work with people from all walks of life. From senior management responsible for multimillion-dollar organizations to new graduates getting ready to open their first small business, we are proud to contribute to our clients' success.

The success of our studio has been built on three specific principles:

- Delivering exemplary customer service while building a strong relationship.
- Providing a quality, state-of-the-art product
- Maximizing a client's return on investment every step of the way

Most clients appreciate what we do and see the obvious benefit of establishing a long-term relationship. Along the way, we have also encountered clients we no longer assist for at least one of three reasons:

They were not nice people. They may have been disrespectful or inappropriate to someone on our management team or crew, or we may have witnessed inappropriate behavior exhibited toward someone else. That is something we just don't want to be associated with.

The core values of our studio did not align with those of the client. C. S. Lewis defined integrity as doing the right thing, even when no one is watching. Doing things right the first time is important to us. We have spent many years building a successful studio and will not jeopardize our success or the success of our individual team members.

They didn't pay their bills. We work extremely hard for each client. Many of our clients do not understand how many clients we work with or how many projects we may be working on at any time. We are constantly recommended to others not only for the quality of our work but for how responsive we are to our clients. If they are not responsive with their checkbook when they receive our invoice and it's not paid per our agreed-upon terms, we will not complete future projects for them.

Sometimes things go wrong. Client issues can create a significant distraction from your normal daily activities. They can also be demotivating to your crew, and if they are not dealt with properly, they can be destructive to any videographer's confidence and business model. It's important to understand that these things can happen, and it may not reflect you, your team, or the services you provide.

If you made a commitment to deliver a product and you are having issues or your client has turned out to be someone who is very different from the person you worked with during the planning process, it's important to put your differences aside and complete the project. It's not only your reputation on the line; if a contract is in place, you may face other legal consequences if you simply walk away. Unless you are in a long-term contract, there is nothing that will require you to complete another project with them.

If your issue is related to billing, depending on how your contract is written, you may be able to slow down the project or even bring it to a halt if your payment terms and conditions are not met. Before you do this, we recommend you seek the advice of your legal counsel.

It doesn't matter whether you're a new or experienced videographer, or whether you're young or old. There are many moving parts to any video production. If you do it long enough, something will not go exactly the way you planned. Do not let your ego or pride stand in the way of getting the job done right. If an issue develops with one of our projects, we typically try to solve it in-house for the first twenty-four hours. If the issue takes more time than that, we make sure the client is brought up to speed and understands not only what happened but also the plan we have in place to adjust the issue. Most clients not only appreciate being kept in the loop, but may offer you additional guidance or direction, which may make a resolution much easier.

Finally, when a problem arises, it needs to be appropriately resolved. Usually there's an additional expense associated with this resolution. It may require you to purchase a new piece of software, provide additional sweat equity, or perhaps bring in an expert for a consultation or even engage them to actively assist with the resolution of your issue. Who covers this expense? Depending on the business model that you have in place, you may have extra funding within your budget that will cover this additional expense. If you're working on a time-and-materials basis, this will impact the bottom line either for you or for your client. We use the following guideline:

- If the issue was a result of something that we did wrong, something that we could have prevented, or a result of our equipment failure, we cover the additional expense.
- If the issue is a result of an action by someone other than us, we will immediately inform the client and have a brief discussion with them. It's essential that everyone understand the issue as well as any anticipated additional expenses required to complete the project. We also make sure everyone understands and agrees who will be responsible for these expenses. If required, we will invoice the client for additional expenses. If the responsibility falls on a party other than the client, it is up to the client to collect any additional fees from other vendors, or the client may choose to deduct it from other invoices that may be due to those vendors.

Discussing problems or issues that may arise during production is a specific area where your communication should be open and direct. An existing relationship and frequent communications with your client will make having a conversation like this easier and more effective. Remember that just like you, clients do not like surprises.

Chapter 10
What's Next?

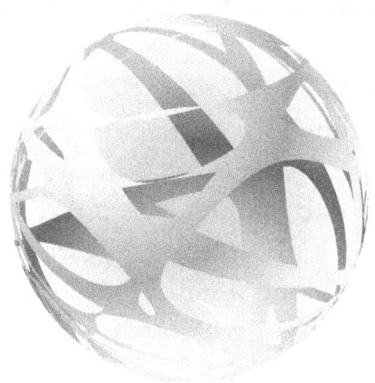

Congratulations! You made it through nine chapters of information representing over ten years of work in our studio and over one hundred years of combined experience within our videography team. We hope the previous chapters provided you with insight into and valuable information about 360-degree video platforms that you can start using today. We trust that the additional general information will also provide you with some guidance or direction as you manage your projects and develop your skills as an amateur or professional videographer. It is our desire that this new information will provide additional assistance to the professional videographer running a business or help the amateur who may be contemplating a business venture.

Our studio is fortunate to be called on to provide a variety of services to many organizations representing numerous market segments and to be a resource for other organizations and individuals providing similar services. I have had the opportunity to speak to individuals and organizations and at trade shows on a variety of topics. It does not matter where I am or who I'm presenting to; I am always asked the following two questions.

What Gear or Software Do You Recommend We Use?

If you go back to the beginning of this book and reread all the way through until this point, you will note that not once did I specifically endorse any camera manufacturer, software vendor, or proprietary piece of equipment. This was done intentionally.

The days of going to a conference and aligning yourself with other participants by which one of the two major camera companies you prefer are long gone. While several reputable big-name companies produce exceptional equipment and software, there are also many companies that are new to the market that are releasing cutting-edge technology at an incredible pace at a fair price point. Whether you're responsible for the acquisition of gear for your organization or to put in your own bag, you need to check out what's happening in the entire market. Sure, there are products that are considered the gold standard, but there are also several products that can deliver what the gold standard can't do. You need to evaluate all your options before buying any piece of equipment or software. Before you select or purchase any new gear or software, reread chapter 3. You'll note at no time in chapter 3 do we suggest that you consider the name of the company or what others around you are using. The equipment you use in the field and computers or editing tools you use back in the studio should be tailored specifically to you and your needs.

Here is a peek at some of the gear we use. If you want to learn about more about the specific gear we use both in and out of the studio, keep an eye on our websites, social media platforms, and YouTube page. Every once in a while, some of our equipment makes it into our photos, videos, or posts.

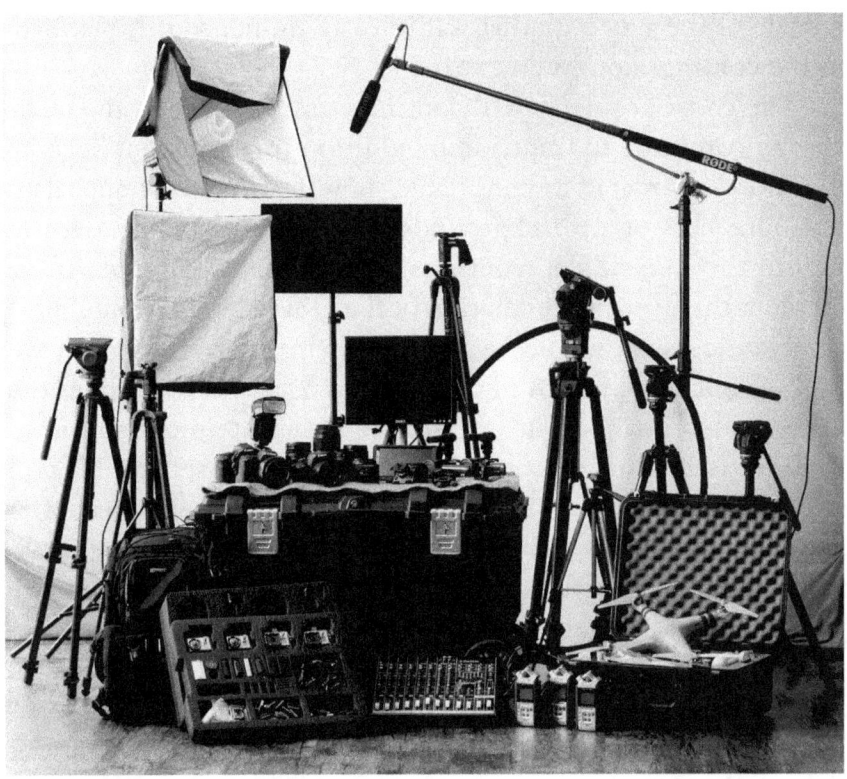

Figure 10.1
Your equipment should be specifically tailored to the needs of your studio.

What's Next?

We love this question! Open-ended questions are awesome. This question is so open ended and the field is changing so fast that there's probably no real answer. However, to wrap up this chapter, let me talk about what I believe is next for 360-degree video. I obviously don't have a crystal ball that will provide us with an in-depth or accurate prediction of the future, but my team and I stay connected with many different sources, both inside and outside of the industry. In my opinion, let me repeat that, in my opinion, I believe it is reasonable to predict the following.

Everyone in the industry agrees that 360-degree video is in a state of infancy. Advances in technology both inside and outside of the video market will greatly influence this platform in the coming months and years.

The cost of cameras will continue to drop even as the technology continues to improve. In addition to equipment becoming more affordable and available, new features will appeal to videographers of all backgrounds and skill levels. The features within the camera will continue to evolve and improve. If we consider the historical evolution of the DSLR and extreme sport camera models, it would be safe to anticipate or predict that we will see not only improved options for compression without the loss of video quality, but we can also anticipate more advanced tools for the management of lighting across the sphere. It's also safe to predict that we will most likely see cameras that offer streaming capability, whether through a wireless or wired network. Finally, as virtual reality continues to go mainstream, we would anticipate that a derivative of the bi-ocular camera that would allow three-dimensional capture may become a standard feature on 360-degree cameras. I'll go out on a limb and say that we might even see these new cameras significantly impact the extreme sport camera market, for the newest releases are offering many of the same features (waterproof, shockproof, compact, etc.) but include the ability to acquire 360-degree video at a similar price point.

Back at the studio, we predict the large players in the software industry will continue to develop their tools to improve the options for editing 360-degree video. Specific tools for advanced editing and management of spherical video as well as previewing it at the desktop level are most likely in our future.

We anticipate the greatest potential area of expansion and improvement will be the distribution platforms for 360-degree video. As this medium becomes more mainstream, we anticipate that social media and proprietary platforms will not only improve their playback interfaces but also improve their upload capabilities and reduce their processing times. As theses distribution channels continue to improve, we also anticipate that many of the software platforms used for editing will improve their distribution interface as they have done in the past for distribution to some of the larger traditional video platforms that are being used today.

Finally, we will soon see new technology that will allow the end user to become further immersed in this platform. It is my prediction that virtual reality goggles or other heads-up display technology will gradually become mainstream; just as with the cameras, advances in the display

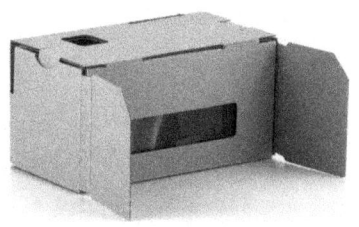

Figure 10.2
Cardboard cutouts will soon be a thing of the past.

technology will allow it to be more affordable and enjoyable to the general public. The use of cell phones with cardboard cutouts will become a thing of the past.

There is no doubt that the integration of this new technology into your offerings and workflow has almost endless possibilities. While I cannot say for certain exactly what this will do for my individual studio offerings in the next ten years, I know it's the start of something big, and I look forward to leading the way.

I wish you much success in your endeavors and invite you to join me on this journey.

Appendix Contents

Equipment Checklist for an Effective Shoot

Videographers of all levels of experience may find the following equipment pack list helpful as they prepare for a shoot. The purpose of this list is to provide you with a list of items you may need for your shoot. We review this list before every shoot to ensure we do not leave the studio without something we may need.

Not every shoot is alike, so not every piece of equipment is required on every shoot. This list should be used as a guide to help you select what you specifically need for a particular shoot.

Cameras
- [ ] camera bodies
- [ ] lenses, lens cleaner, and cloth
- [ ] batteries, charger(s), AC power cords
- [ ] memory cards
- [ ] tripods and special mounts
- [ ] remote controls/corded actuators
- [ ] follow focus rig
- [ ] hoodman

Sport Cameras
- [ ] cameras
- [ ] batteries
- [ ] memory cards
- [ ] tape and specialty mounts
- [ ] waterproof housings and anti-fog strips

360-Degree Cameras
- [ ] camera housings
- [ ] batteries
- [ ] memory cards
- [ ] tape and specialty mounts
- [ ] light stands
- [ ] remote controls
- [ ] live preview device

Lighting
- [ ] lightboxes
- [ ] umbrellas
- [ ] lights stands
- [ ] various bulbs (and extra bulbs)
- [ ] incandescent
- [ ] compact fluorescent
- [ ] LED
- [ ] reflectors and Stands
- [ ] dimmer control boxes
- [ ] extension cords
- [ ] dimmer box

Data Management
- ☐ laptop, tablet
- ☐ batteries
- ☐ memory cards
- ☐ data storage devices
- ☐ portable hard drives

Audio
- ☐ microphones
- ☐ wireless transmitters and receivers
- ☐ batteries
- ☐ boom poles and stands
- ☐ audio recorder
- ☐ memory cards
- ☐ audio cables
- ☐ headsets
- ☐ mixing board
- ☐ Data Management
- ☐ laptop, Tablet
- ☐ batteries
- ☐ memory cards
- ☐ data storage devices
- ☐ portable hard drives

Monitors
- ☐ director's monitor
- ☐ teleprompter's monitor
- ☐ teleprompter's head
- ☐ wireless transmitters and receivers
- ☐ monitor stands

Backdrops
- ☐ stands and cross bars
- ☐ green screen
- ☐ backdrop cloths
- ☐ spring clamps
- ☐ weights

Stabilizer Systems
- ☐ glide cam
- ☐ steady cam micro
- ☐ steady cam standard
- ☐ steady cam vest
- ☐ spring tool

Drone
- ☐ aircraft
- ☐ aircraft remote control
- ☐ camera remote control
- ☐ batteries
- ☐ memory cards
- ☐ video transmitter and monitors
- ☐ first-person view components
- ☐ training tether
- ☐ extra blades
- ☐ FAA log book
- ☐ flight plan

Crane
- ☐ crane frame
- ☐ tripod and cart
- ☐ counterweights
- ☐ remote monitors and batteries
- ☐ camera remotes
- ☐ chains and cables
- ☐ tool bag

Tracking System
- ☐ tracking dolly
- ☐ track and connectors
- ☐ cleaning cloth
- ☐ lubricant

Safety Equipment
- ☐ safety vests
- ☐ helmets
- ☐ boots
- ☐ earplugs
- ☐ rain gear

Miscellaneous
- ☐ tape measure
- ☐ multitool extension cords
- ☐ duct tape
- ☐ clear packing tape
- ☐ paper towels
- ☐ all-purpose spray cleaner
- ☐ glass cleaner

Appendix B
Planning Checklist for an Effective Shoot

Videographers of all levels of experience may find the following planning checklist helpful as they prepare for a shoot. Not every shoot is alike. This should be used as a guide to help you consider what you may need to do for a specific shoot. It is not intended to be a step-by-step instruction sheet.

Proposal/Contracting
- ☐ define deliverables
- ☐ define responsibilities
- ☐ set budget
- ☐ define payment terms

Scripting
- ☐ set story line
- ☐ secure talent
- ☐ choose location
  - ☐ permits (if needed)
  - ☐ permissions (if needed)
- ☐ obtain releases (for talent/property)

Audio
- ☐ complete audio plan
- ☐ define voice-over
  - ☐ script
  - ☐ talent
- ☐ choose soundtrack

Data Management
- ☐ develop data plan for storage, transfers, and backup

Operations
- ☐ set up equipment—primary and backup cameras
- ☐ confirm the following:
  - ☐ batteries charged
  - ☐ spare batteries available
  - ☐ empty memory cards in gear
  - ☐ spare memory cards available
  - ☐ lighting in place—camera set for lighting
- ☐ shoot sample video or complete live preview of sphere
- ☐ record sample audio and confirm appropriate levels on playback
- ☐ establish gear cache and backup plans
- ☐ assume nothing (be alert and prepared for anything)
- ☐ once in place, confirm power, memory cards, and lighting; complete test video or perform live preview prior to shoot
- ☐ during the shoot, confirm your gear is recording quality video and audio

Appendix C
Sample Postproduction Workflow Diagrams

Amateur and professional videographers alike will benefit from a defined workflow for postproduction. A workflow will not only keep your production organized but will result in a more efficient management of your projects. This efficiency may result in your saving time, improving your profit margin, or perhaps delivering your final production ahead of schedule.

While every videographer or studio should adopt a standard workflow, not every workflow will work for everyone or for every project. A successful workflow is nothing more than an outline. It is a flexible process that should guide the videographer or editor through the complex process of postproduction while providing them the ability to slightly deviate as required for specific projects or productions.

We have partnered with studios who keep a laminated copy of their workflow on the desks within an editing suite and others who print out the workflow and use it as a checklist as the project moves forward. At our studio, we take it one step further: In addition to using a printed copy, we also annotate any changes or deviations along the way, processes skipped, or special processes included with the production. This document is used not only during the editing of the specific project but also retained for future reference. In addition to it providing value should we have to recreate the edit in the event of a data loss or should we wish to recreate it later or replicate an effect on another project, we can simply pull the workflow document for quick reference. Eliminating the time to reverse engineer a project later has proven to be a useful benefit several times.

We understand that, like our rather complex planning process, some experienced videographers or editors may be resistant to change and feel this documentation adds an unnecessary step in their overall workflow or project management. Nevertheless, we encourage you to consider this integration because it will add value to your projects and to your offerings over an extended period of time.

Data Management

The following diagram illustrates a sample postproduction workflow for data management.

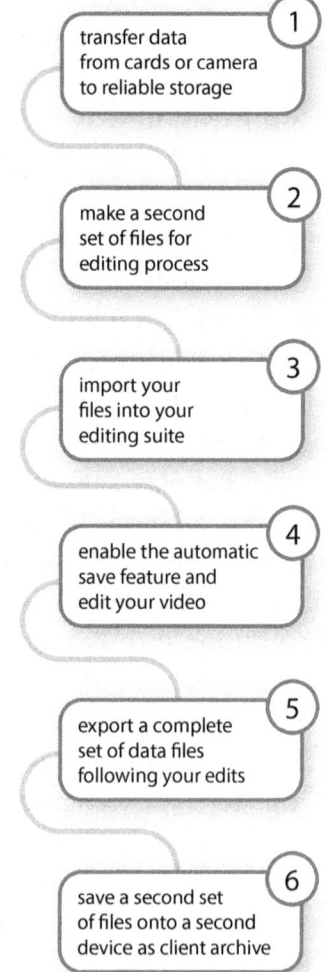

transfer data from cards or camera to reliable storage — 1

make a second set of files for editing process — 2

import your files into your editing suite — 3

enable the automatic save feature and edit your video — 4

export a complete set of data files following your edits — 5

save a second set of files onto a second device as client archive — 6

Figure C.1 Data Management Workflow Diagram

The following is a brief description of each step in the previous diagram.

1. Transfer your data from your memory cards or camera(s) to your storage device. (Make sure this computer is backed up on a regular basis.)
2. Make a second set of files to be used in the editing process.
3. Import your files into your editing suite.
4. Edit your video. (If the software has an auto-save feature, make sure it is enabled.)
5. Export a complete set of data files following your credit.

Make a second set of files that should be stored on a second device as part of a client archive.

138

General Postproduction

The following diagram illustrates a sample general postproduction workflow.

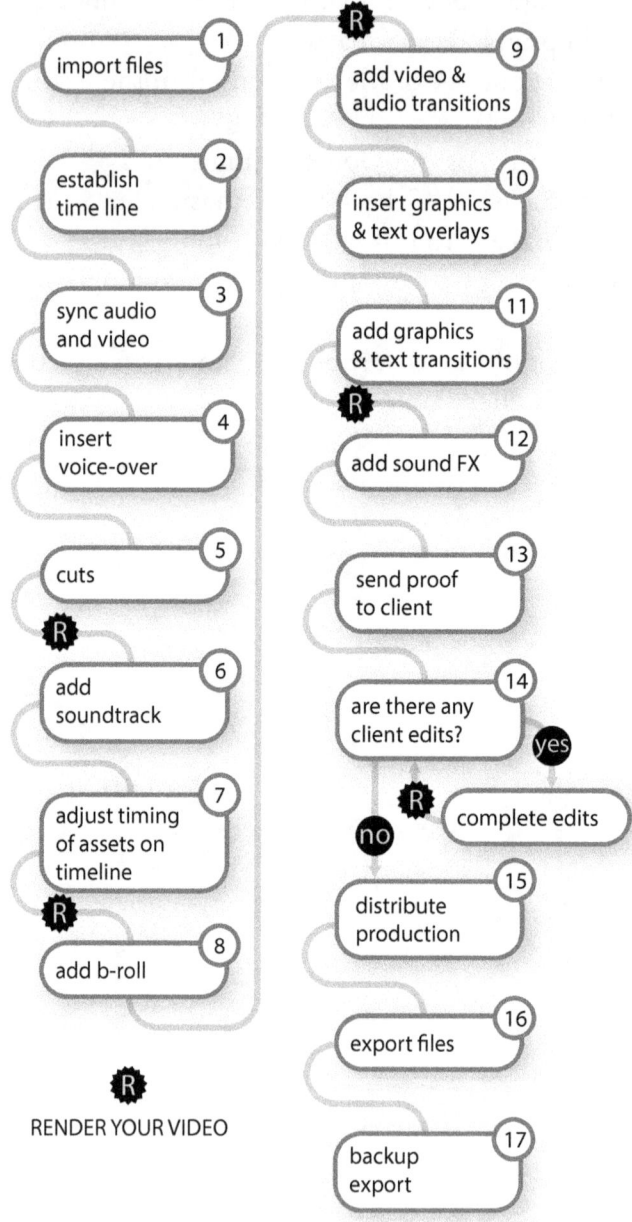

Figure C.2 General Postproduction Workflow Diagram

The following is a brief description of each step in the previous diagram.

1. Import the files from your data source into your editing software.
2. Place all your video and audio files on the timeline.
3. Synchronize all audio and video files.
4. Place your voice-over on the timeline and adjust the timing of your audio and video files with your voice-over.
5. Cut your audio and video files to match your timeline. (Render your timeline.)
6. Add your soundtrack or any background music.
7. Review the assets on your timeline and adjust the individual timing as necessary. (Render your timeline.)
8. Add B-roll footage as necessary. (Render your timeline.)
9. Add any video and audio transitions between your cuts as appropriate.
10. Insert any graphics or text overlays.
11. Add any graphic and text transitions as required. (Render your timeline.)
12. Add sound effects.
13. Send proof to client.
14. Are there any edits returned from the client? Complete the edits. (Render your timeline.) Resend the proof to the client. Repeat this process until there are no additional edits.
15. Distribute your production.
16. Export your files from your editing suite to your client files.
17. Back up your export.

Appendix D

Determining Operational Distances, Stitch Points, and Blind Spots

It is essential for videographers to determine optimal distances, stitch points, and blind spots for any 360-degree camera they intend to use on a shoot. Ideally, they should determine these before a scheduled shoot. We recommend that specific measurements be taken during this determination. Although camera settings may change a little on each subsequent shoot, these measurements will provide you a solid starting point to help you plan and set up your shoot.

You should be familiar with three operational distances:

- **Inner working boundary.** The distance from the camera lens to the point at which your subject is no longer blurry or distorted within the camera's field of view.
- **Outer working boundary.** The distance from the camera lens to the point at which your subject becomes distorted, or the subject of the video becomes so small that the video does not have the desired level of detail. This is a very subjective distance and may vary from project to project. As a guideline, we use the distance at which we cannot clearly distinguish the difference in color between the colored part (iris) and the white part (sclera) of the eye.
- **Sweet spot.** The average distance between the inner and outer working boundary (or halfway between the inner and outer working boundary).

Although it is easy to visualize this on a flat diagram, remember that 360-degree video is all about working in the sphere. It may be helpful to visualize the camera inside a series of balls. The camera is inside a ping-pong ball, which represents an inner boundary. The ping-pong ball is within a tennis ball, which represents the sweet spot. The tennis ball is within a softball which represents the outer boundary. If you are aligning objects to be in the sweet spot, they would be anywhere on the covering of the tennis ball.

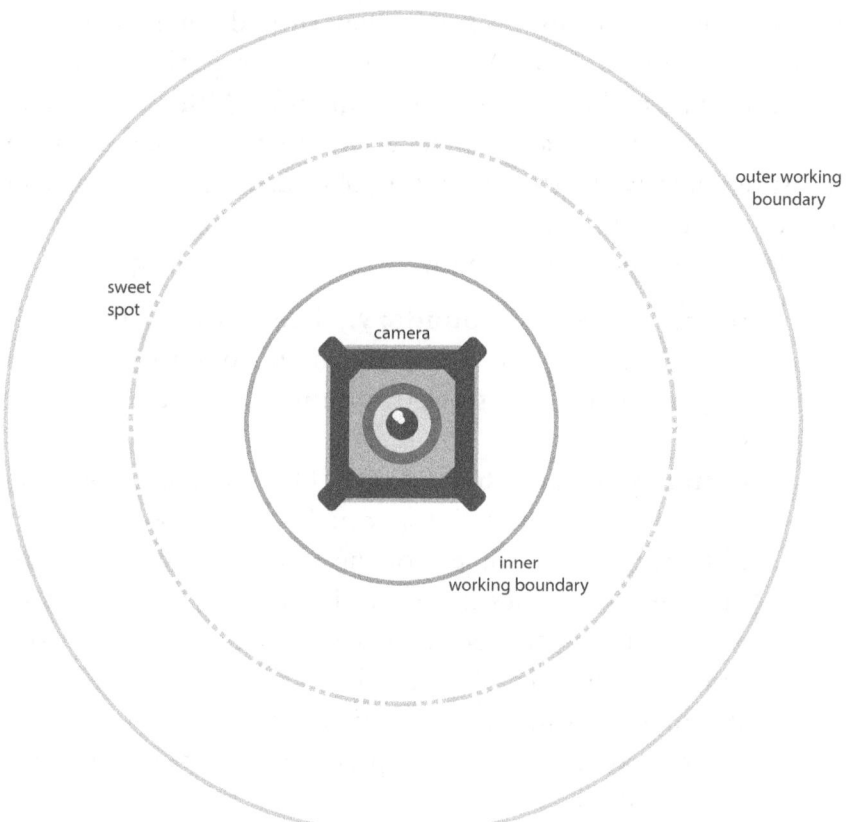

Figure D.1 Proper Camera Placement

For proper camera placement, it's important that you understand your camera's working boundaries and sweet spot.

How to Determine Optimal Distances for Subject Matter

Place your camera on a mount that can be adjusted vertically.

1. Set the camera to one half the height of the room you plan to use for determining these distances.
2. Turn your camera on and activate its live preview mode.*
3. Put your face directly in front of the primary forward-facing lens while viewing the live preview on your connected preview device. You should see some distortion or aberrations. Slowly back away from your camera until the distortion is alleviated. With most cameras, you will see noticeable improvement at about eighteen inches from the lens and clear video anywhere from two to three feet from the lens. Measure the distance from the lens to the point where the distortion is alleviated. This is the inner working boundary.
4. Make sure you are still in the live preview mode. Continue to back away from the lens until one of two things happens: your image becomes distorted, or the subject of the video becomes so small that the video does not have the desired level of detail. Measure this distance, which is the outer working boundary.
5. Now take the average of these two measurements or find the midpoint between the inner and outer working boundary. This distance is generally considered the sweet spot for your camera.

* If your device does not have a live preview mode, you can find these dimensions by slowly repeating the process of recording video and playing it back.

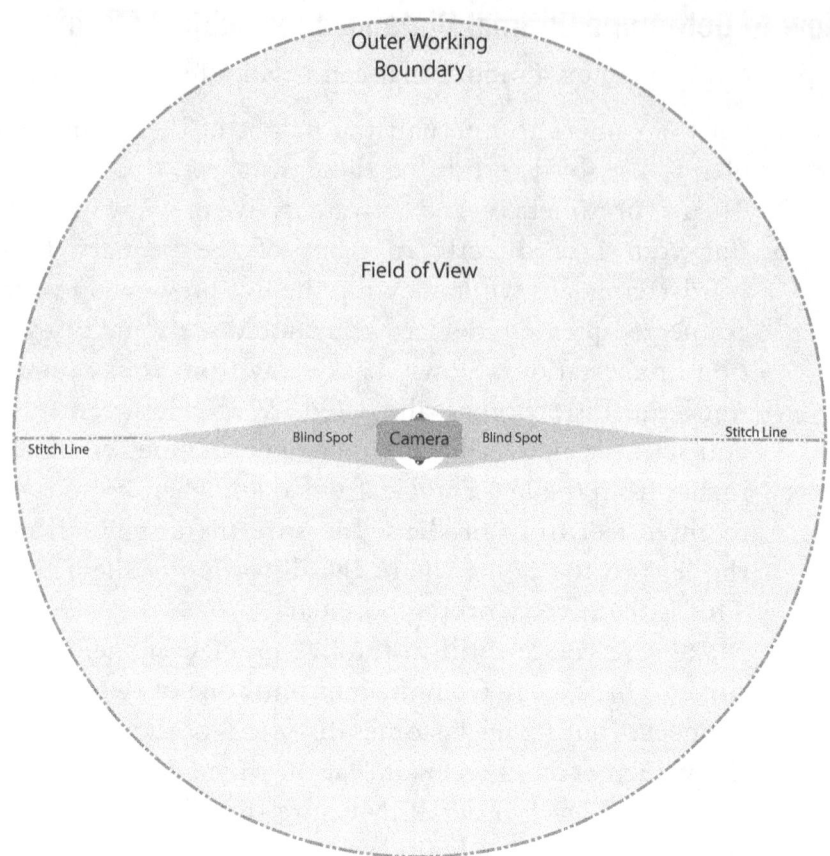

Figure D.2 Camera Blindspots

Experienced videographers will understand and use a camera's blind spots to their advantage.

How to Determine Your Camera's Stitch Points and Blind Spots

A blind spot is an area that the camera's lenses are unable to pick up.

A stitch point or stitch line is the point at which the fields of view of two or more lenses overlap and the image is stitched together either on the camera or by the software in postproduction. In this procedure for determining stitch points, we are assuming that the videographer is using a single camera with multiple lenses and that the stitching is occurring within the camera.

You will want to pay very close attention to how you complete the next few steps so you can replicate the procedure repeatedly.

1. Make sure your camera is on and in live preview mode.★
2. Take two fingers and place them directly in front of the primary lens. You want to stay within an inch (a finger's width) of the camera's lens as you complete steps 3 and 4.
3. Slowly move your hand to the right and keep your eye on the preview device. Depending on your specific camera/lens configuration, you may see your fingers disappear anywhere from 60 degrees to 90 degrees from the center point of the lens. Remember, you are working in a sphere, so that a blind spot will be circumferential, like the working boundaries and sweet spot.
4. As soon as you find the blind spot, stop moving your hand and record the approximate angle.
5. Find your blind spot again with your fingers within an inch from the lens. Slowly move your fingers away from the lens in a straight line away from the camera at the approximate angle where the blind spot was detected.

★ If your device does not have a live preview mode, you can find the blind spots and stitch lines by slowly repeating the process of recording video and playing it back.

As you get farther from the camera, the blind spot will dissolve into video footage. This is the stitch line.

6. Next repeat steps 3–5, except move your hand to the left instead of the right. You should be able to find the blind spot and corresponding stitch line.

7. Finally, if you have more than two lenses, repeat these steps for each lens and note the overlapping regions of the blind spots.

Appendix E
Using Blind Spots to Your Advantage

Blind spots with 360-degree video cameras are common. Depending on the positioning of the lenses and the field of view of each, there is usually an area close to the camera that none of the lenses will cover. Different platforms will have different configurations and thus will have different blind spots. In our review of the different cameras, we found that cameras with a larger number of lenses generally had smaller blind spots, as there were additional areas of overlap for each field of view. In some instances, we found cameras that had more lenses also had more blind spots, but the area of the blind spots was smaller. We also noted that as we increased the number of lenses, we had additional stitch lines, which could present additional challenges, depending on the subject.

After all our evaluations and research, our favorite camera had two lenses and two distinct blind spots. We used this camera to complete most of our experimental shooting as well as to create the illustrations for this book. We found this configuration to be beneficial for a few reasons.

Although the blind spot is technically circumferential around the lens, for our setup we identified it on a flat plane and referred to it as left or right of the primary camera. We used these two spots as follows.

- **Camera redundancy.** As we have mentioned in a few places in this book, we believe there is an advantage in using multiple cameras on every shoot. The blind spots for this camera let us take two cameras and place them side-by-side. Neither camera can see the other camera. We typically put the second camera to the right of the

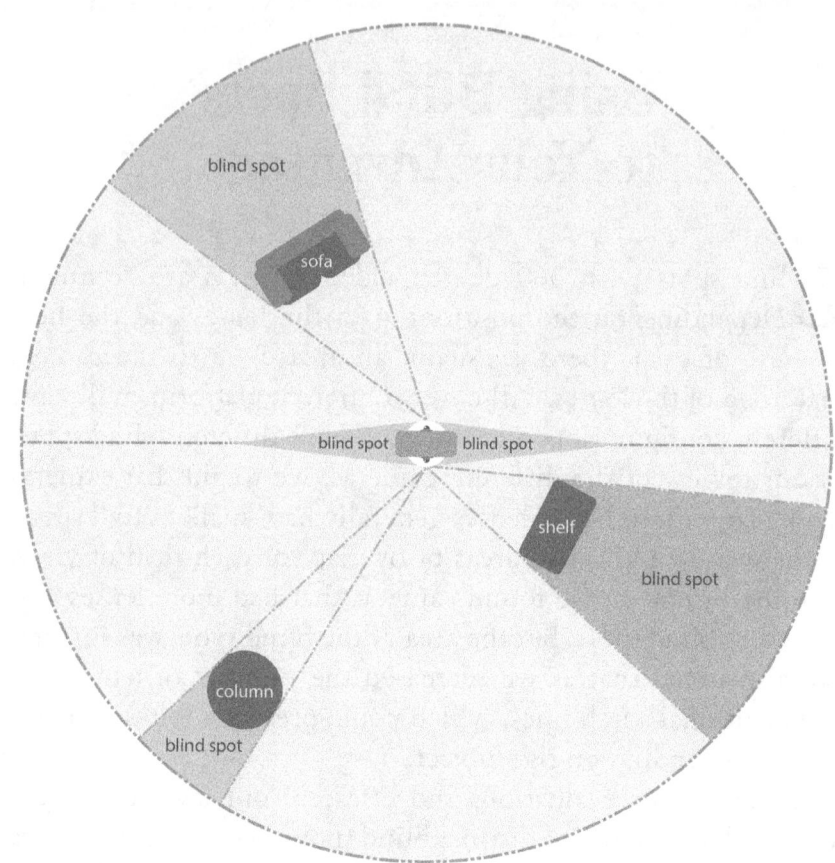

Figure E.1 Blindspots as an Advantage

Blind spots may be used to hide other cameras, gear, or crew members.

primary camera. It could also be placed to the left; it really does not matter. What is important is that you identify which is the primary camera and which is secondary so you can keep the data cards and the transferred data straight. If you don't keep the data separated and start pulling data from different cameras onto your editing timeline, you will end up with a jump in your focal point.

- **Audio acquisition.** For our simple 360-degree shoots, we use a wireless lavalier microphone that transmits to a receiver attached to a digital recorder. Our gear is small enough that the receiver and recorder can be placed within the blind spot to the left of the camera.

If you've done this before or are using this book as a guide to set up your shoot, you are probably scratching your head and wondering, "How could they do that?" We developed a custom bracket that attaches to the top of a light stand to hold this gear together. Although I cannot show you the actual piece of equipment at this point, we hope it will be available for purchase in the coming months.

As we discussed in chapter 5, in addition to the blind spots of a camera, blind spots also occur because of objects within the sphere. We have used these blind spots to hide our equipment cache, teleprompters, and mixing gear. We've hidden crew members, clients, and even objects on the set that the client did not want displayed.

Blind spots are valuable spaces in the sphere that are often overlooked. To use them successfully, you must spend some additional time previewing the sphere before recording begins.

Index

Note: an *f* indicates a figure.

About the Author

John Hussar is the founder and managing partner of Grey Goose Graphics, a multimedia advertising agency in upstate New York. His pharmaceutical marketing experience combined with over twenty years as an emergency service manager enables him to relate to and work with a diverse client base. 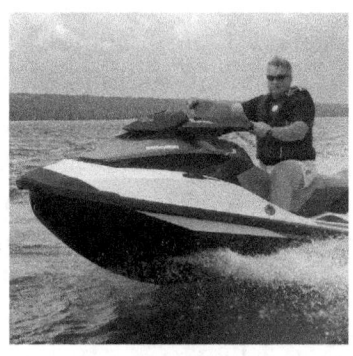 John's unique management style and the ability to "triage" life's daily events has facilitated his personal success and has led to Grey Goose Graphics' consistent growth year after year.

Early in his career he developed a love for video and has been an early adopter of technology. On a regular basis, John leads and participates in several think tanks that focus on leveraging the power of video and integrating it with digital and social media platforms.

With a strong background working within and representing clients from nonprofit organizations, John has developed many products and solutions specific to this market, allowing these organizations to effectively leverage their often-limited funding to create successful marketing campaigns and products.

When not in the studio on or on the road, John remains an active in his community as a volunteer firefighter and serves on the board of a local nonprofit organization. He enjoys the outdoors, camping, and jet skiing with his wife MaryLou and his daughter Alexis.

Contact John at Author@SphericalVideoBook.com.

About Grey Goose Graphics

Grey Goose Graphics LLC is a small, dedicated, and client-focused studio located in the Southern Tier of Upstate New York, which was established in 2006 by John Hussar. Comprised of a cohesive team of experienced professionals who take pride in exceeding our customers' expectations, Grey Goose Graphics provides multimedia production services to a diverse international client base for numerous educational, documentary, promotional, and entertainment productions.

In addition to a dedicated video production division, we specialize in the creation of custom solutions and products in the areas of new website development, existing website maintenance, web hosting management, company identity and branding, graphic design, print and promotional product services, email marketing, and social media management.

Our project team leaders have extensive experience and have repeatedly demonstrated their strong business backgrounds, ability to multitask on your project, and meet deadlines. Our project team leaders will keep you informed of your project's progress so you will be connected with your project from start to finish.

We recognize demonstrating a return on your investment is essential, and we take this responsibility seriously. We are not satisfied until you are.

We consistently bring value to our clients in many ways by exceeding expectations. If your current agency is not exceeding yours, you should be talking with us.

Visit us on line at www.GreyGooseGraphics.com to learn more or to connect with a member of our team.